The Cost of Climate Policy

The Sustainability and the Environment series provides a comprehensive, independent, and critical evaluation of environmental and sustainability issues affecting Canada and the world today.

1 Anthony Scott, John Robinson, and David Cohen, eds., *Managing Natural Resources in British Columbia: Markets, Regulations, and Sustainable Development* (1995)

2 John B. Robinson, *Life in 2030: Exploring a Sustainable Future for Canada* (1996)

3 Ann Dale and John B. Robinson, eds., *Achieving Sustainable Development* (1996)

4 John T. Pierce and Ann Dale, eds., *Communities, Development, and Sustainability across Canada* (1999)

5 Robert F. Woollard and Aleck Ostry, eds., *Fatal Consumption: Rethinking Sustainable Development* (2000)

6 Ann Dale, *At the Edge: Sustainable Development in the 21st Century* (2001)

7 Mark Jaccard, John Nyboer, and Bryn Sadownik, *The Cost of Climate Policy* (2002)

SUSTAINABILITY AND THE ENVIRONMENT

Mark Jaccard, John Nyboer, and Bryn Sadownik

The Cost of Climate Policy

UBCPress · Vancouver · Toronto

09 08 07 06 05 04 03 02 5 4 3 2

Printed in Canada on acid-free paper that is 100% post-consumer recycled, processed chlorine-free, and printed with vegetable-based, low-VOC inks.

National Library of Canada Cataloguing in Publication Data

Jaccard, Mark Kenneth.
The cost of climate policy

(Sustainability and the environment, ISSN 1196-8575)
Includes bibliographical references and index.
ISBN 0-7748-0950-7 (bound); ISBN 0-7748-0951-5 (pbk.)

1. Greenhouse gas mitigation – Economic aspects – Canada. 2. Environmental policy – Economic aspects – Canada. 3. Environmental protection – Economic aspects – Canada. I. Nyboer, John, 1954- II. Sadownik, Bryn, 1972- III. Title. IV. Series.

HC120.E5J32 2002 363.738'746'0971 C2002-910516-1

Canada

UBC Press gratefully acknowledges the financial support for our publishing program of the Government of Canada through the Book Publishing Industry Development Program (BPIDP), and of the Canada Council for the Arts, and the British Columbia Arts Council.

This book has been published with the help of a grant from the Humanities and Social Sciences Federation of Canada, using funds provided by the Social Sciences and Humanities Research Council of Canada.

Set in Stone by Artegraphica Design Co. Ltd.
Printed and bound in Canada by Friesens
Copy editor: Dallas Harrison
Proofreader: Deborah Kerr
Indexer: Annette Lorek

UBC Press
The University of British Columbia
2029 West Mall
Vancouver, BC V6T 1Z2
604-822-5959 / Fax: 604-822-6083
www.ubcpress.ca

Dedicated to

Doris and Lou Jaccard

Cindy, Michelle, Beth, and Debbie Nyboer

Gertrude Osborne

For updates and further information on the issues discussed in this book, visit the Energy and Materials Research Group Web site: <www.emrg.sfu.ca/costofclimatepolicy>.

Contents

Figures and Tables

Figures

Tables

Acronyms

AEEI	autonomous energy efficiency index
AMG	Analysis and Modelling Group
AQVM	Air Quality Valuation Model
CaSGEM	Canadian Sectoral General Equilibrium Model
CEM	community energy management
CEOU	*Canada Emissions Outlook: An Update*
CFCs	chlorofluorocarbons
CGE	computable general equilibrium
CH_4	methane
CO	carbon monoxide
CO_2	carbon dioxide
DET	domestic emission trading
E/Q	energy intensity
ESUB	elasticities of substitution
GDP	gross domestic product
GHG/E	GHG intensity of energy
GHGs	greenhouse gases
H_2	hydrogen
HCFCs	hydro-chlorofluorocarbons
HFC	hydro-fluorocarbons
HVAC	heating, ventilation, and air conditioning
IPCC	Intergovernmental Panel on Climate Change
N_2O	nitrous oxide
NCCP	National Climate Change Process
NEMS	National Energy Modeling System
NIS	National Implementation Strategy
NO_X	nitrogen oxide
O_3	ozone
OECD	Organization for Economic Cooperation and Development
PFC	perfluorocarbons

RPS	renewable portfolio standard
SF_6	sulphur hexafluoride
SO_2	sulphur dioxide
VES	vehicle emission standard
VOCs	volatile organic compounds

Preface

Public policy often – some would say too often – gets reduced to the question "What does it cost?" In the eyes of many, the big challenge of climate change policy is that we may incur large costs in the near and medium term for uncertain benefits in the distant future. The cost question looms large in this kind of dilemma. Policy makers need to know as much as possible about the costs of taking action now to reduce greenhouse gas (GHG) emissions. The media and public need to know too.

Yet surprisingly little research has been conducted on the costs of reducing emissions, both on an international level and specific to Canada. Of the studies that have been conducted, most are linked to different interest groups who start from a particular perspective and then consistently find results that support their position. Even though some of these studies may be well designed and relatively objective, it is hard for independent observers not to be skeptical. This is not the only concern with cost estimates. Researchers in this area are bedevilled by a methodological schism that has some focusing on the large technological potential for reduced emissions and others focusing on the inertia to change resulting from the current preferences of businesses and consumers for specific technologies and ways of doing things.

As members of the Energy and Materials Research Group in the School of Resource and Environmental Management at Simon Fraser University, we have been engaged in estimating the costs of GHG policy for sixteen years. We have conducted this research independently but also as consultants to government, industry, and nongovernmental organizations. We are proponents of an approach to cost estimation that bridges the methodological schism in our field. We explicitly model technologies, as do those who focus on the large technological potential. But we also explicitly incorporate the best available information on the preferences of businesses and consumers for different technologies and for different goods and services from the economy. With this hybrid approach, we try to help policy makers and others see how cost estimates change as we adjust our assumptions about technological innovation and the potential for change in business and consumer preferences. This approach is intimately tied to the policy-making function because the effectiveness of policies in this area depends on their ability to navigate the critical interplay between technological innovations and preferences.

Concern over this recurrent methodological split motivated us to write this book. While our focus is on a single country, the book's relevance is much broader. The international community of energy/environment policy analysts will be interested in our approach to addressing the major methodological dispute about costing GHG abatement. Hybrid approaches, with both technological explicitness and preference incorporation, are only beginning to emerge in the theoretical and applied literature. These analysts will be especially interested in Chapters 2, 3, 4, 7, and 8.

At the same time, the methodological approach that we espouse is best understood in a concrete application to a specific region, country, or group of countries. For concerned Canadians (public servants, politicians, industry, nongovernmental organizations, media, environmentalists, academics, regional interests) who want to know more about their country's GHG abatement costs and policy options, all chapters of the book will be of interest. For this reason, we present the issues and methodology in Chapters 2 and 3 in a nonmathematical format that will be accessible to many readers who are otherwise neophytes in the world of costing GHG emission reductions. In Chapter 4, we show what these costs mean in terms of the daily energy prices, technologies, urban form, land use management, and lifestyle choices facing Canadians. Our detailed analysis is disaggregated by sector in Chapter 5 and by region in Chapter 6. While Canadians have an interest in all of these results, some readers will want to focus on the sector or region that most concerns them. However, in the interest of brevity, we have omitted the most detailed results, so we advise readers who want more specific information to visit our Web site and contact us directly (http://www.emrg.sfu.ca). For those readers who are newer to the issue, Chapter 1 provides a brief overview of where GHG abatement fits within the context of the climate change threat and its response options.

The relationship between cost estimation and policy choice is critical in our view. This position comes out especially in Chapters 2, 7, and 8, in which we explain our definition of costs (itself controversial because of the methodological schism) and what our estimates of costs mean for policy choices. In particular, while we support the current emphasis in government policy on information, voluntarism, and partnerships, we are skeptical, given our cost estimates, of the ability of this approach alone to incite the profound changes in technology and preference required to significantly reduce emissions. We think that there is merit in experimenting today with policies that are especially designed to stimulate long-run technological innovation and long-run shifts in the preferences of businesses and consumers. Moreover, such policies can be designed in ways that cause minimal cost and political resistance in the short term while helping to lower both financial cost and other costs related to consumer preferences over the long term.

Although we focus our analysis on Canada's GHG emission target from the international Kyoto Protocol of 1997, the ultimate timing and magnitude of Canada's GHG emission reduction effort are likely to change, perhaps several times, in the coming years. For this reason, we present a cost curve in Chapter 4 that shows what different levels of emission reductions might cost. Because this curve is highly uncertain, given the very uncertainties in technology and preference that are the rationale for our approach, we conduct sensitivity analysis and devote portions of the final two chapters to the formulation of policy approaches that can help to accelerate the generation of information that will ultimately reduce some of these uncertainties.

In this book, we draw on the analyses of other works that have parallelled and supported our own estimates of national costs and from the synthesizing work of many key people both inside and outside government. Thanks are due to these individuals, too numerous to name here. Specifically, we wish to thank former and current members of the Energy and Materials Research Group who worked hard over several years in developing our model and calculating these and other cost estimates: Alison Bailie, Chris Bataille, Roberto D'Abate, Alison Laurin, Rose Murphy, Mallika Nanduri, and Amy Taylor.

We have created a Web site <http://www.emrg.sfu.ca/costofclimatepolicy> that complements, extends, and updates the content of *The Cost of Climate Policy*. The site contains information on international negotiations, national processes and commitments, updated cost estimates, studies by ourselves and others, and links to other key Web sites. This site will be updated on a regular basis.

Introduction

It is not a question of whether the Earth's climate will change, but rather by how much, how fast and where.

– Robert T. Watson, chairman of the Intergovernmental Panel on Climate Change, address to the Sixth Conference of the Parties of the United Nations Framework Convention on Climate Change, 20 November 2000.

That's a big no. The President believes that it's an American way of life, and that it should be the goal of policy makers to protect the American way of life. The American way of life is a blessed one.

– White House press secretary stating US president George W. Bush's position on whether Americans need to change their lifestyles to reduce energy consumption, 8 May 2001.[1]

Defining the Issue

The prospect of human-induced climate change is arguably one of the greatest and most controversial challenges that humankind has ever faced. Many scientists contend that over the past few hundred years consumption of fossil fuels by the world's expanding human population, and other human-induced changes to the biosphere from forestry and agriculture, are increasing atmospheric concentrations of greenhouse gases (GHGs), with potentially devastating impacts. These trace gases in our atmosphere are crucial to maintaining the temperature level of the Earth, and impacts, although hard to predict, may include a rise in sea level, desertification, more extreme weather events, and a loss of biodiversity because of the rapidity of change.

For more than a decade, the world has been grappling with this issue. Opinions such as that voiced in the White House quotation above are not uncommon. Taking proactive action is difficult when it involves changes to current lifestyles and consumer preferences and when a high level of cooperation is needed from widely disparate and unequal nation-states. In contrast to reactive action, proactive action cannot see the consequences before the world puts out money and effort. Some argue that climate change is part of natural variation and is basically unalterable. Others suggest that, if climate change were true, then the effects will not be harmful, and the Earth's habitability will actually improve. What if great effort and expenditure to mitigate climate change are shown to be futile or the threat of negative impacts proves to be false? On the other hand, what if the threat is real, but we do nothing in the false belief that reducing emissions will be too costly? Is the life-sustaining capacity of the Earth at risk? What might action or inaction mean for our children and us?

Most Canadians share the international concern over global warming, and Canada has participated in all international meetings on climate change and global warming, including the 1997 Kyoto Conference of Parties and its resulting protocol.[2] In this protocol, the Canadian government agreed to a target for Canada to reduce GHG emissions to 6% less than 1990 levels by 2008-12. Since that time, Canada has been under pressure both to move forward with specific strategies for making reductions and to prepare for further international negotiations. Future discussions must work out contentious issues regarding who must make reductions and how they can be made, as part of either the Kyoto Protocol or a new negotiated agreement. Perceptions about the economic cost of reducing emissions have guided negotiating positions and the willingness of countries to move ahead with GHG mitigation policies. Many analysts and environmental advocates argue that emission reductions will not be costly, at least for achieving the Kyoto targets. Yet, in stating that the United States will not adhere to the Kyoto Protocol, US president George W. Bush cited economic costs and increased risks of higher unemployment.[3] Many analysts and industry representatives agree with this position.

When Canada signed the Kyoto agreement, the federal government had done little to estimate the costs of a concerted national effort to reduce emissions. There had been various studies by academic researchers, consultants, think tanks, and interest groups, resulting in an array of estimates that ranged from net economic benefits to the loss of several years of economic growth. Neither the federal government nor the provincial and territorial governments had commissioned more than partial studies on the costs of reducing GHG emissions. Even so, the government of Canada signed the Kyoto Protocol and committed to about a 25%-30% reduction from what emissions would otherwise be in 2010 without the government or Canadians having a clear idea what this commitment might cost them.[4]

Indeed, while many of the media and informed public appeared to support the government's position, members of the business community and some provincial politicians expressed shock that the government would make such a commitment without knowing its cost, and they warned that the cost impacts of the accord would be devastating. Tom d'Aquino, president of the Business Council on National Issues, stated that the Kyoto Protocol sets an unrealistic target and timetable that cannot be met without significant economic cost.[5] Premier Ralph Klein has warned that the agreement could cost Alberta's economy "trillions of dollars" and has likened the commitment to the cost of the National Energy Program to Alberta: "I still have to remind people of the literally thousands and thousands of people who lost their jobs, their homes and their dignity. Do we want to see that kind of thing again? No."[6]

Our goal in this book is to provide Canadians with an independent and comprehensive estimate of what it might cost to reduce GHG emissions by significant levels from what they would otherwise be in the year 2010, the Kyoto target date. We use the national target established at Kyoto as the basis for our detailed cost estimates, breaking them down by region and sector of the economy. However, we also show how the costs would differ at alternative target levels. While the Kyoto target provides a useful reference point, the general relationship between reduction levels and costs will be a critical decision-making consideration over the coming years and even decades as targets are explored and adjusted in the face of changing scientific, technological, and international conditions. Thus, our estimates include both a sensitivity analysis of targets lower than that of the Kyoto agreement and the development of a national cost curve to show what higher targets might imply. As long as it is acknowledged that Canada must eventually act to reduce its GHG emissions, our results will be of interest to Canadians.

Since 1986 our research unit at Simon Fraser University – the Energy and Materials Research Group in the School of Resource and Environmental Management – has analyzed strategies and policies that lead to a more sustainable flow of energy and materials in society. The cost estimates presented in this book are based on original analyses conducted by our research group to assist Canada's national process for formulating GHG abatement policy. Although the results that we present here were primarily generated during that process, we have put our own focus on these results and have conducted subsequent research that forms part of the body of work presented here.[7]

Analysts try to support political decision making by providing reasonable and unambiguous cost estimates. This is understandable. But cost estimates are not unambiguous. They are intimately linked to our differing perceptions about the economy, about whether consumers have free choice or are manipulated, about whether market outcomes are a better indication of costs than analysis by engineers or economists, and about whether markets are efficient or not. We explore and explain the key dimensions of this issue, both generally and specifically as they apply to our cost estimates. We explain the critical methodological issues that lead to divergent cost estimates and provide our views on the policy implications of our research. Governments have talked about reducing GHG emissions but have said little about specific policies and what they might mean and cost. Does the choice of policy have an effect on cost? On public acceptance? On who bears the greatest share of the costs?

Cost estimates are also meaningless to most people unless they see them in terms of the prices and technologies that they face on a daily basis. Thus far no one has provided this kind of detail for Canadians. What do the costs of reducing GHG emissions mean for the price of gasoline? What do they mean for natural gas and electric utility bills? What is the implication for the costs of acquiring and using a car? What do these costs mean for the future of our communities? Will some industries collapse, some communities and regions decline? To address these questions, we present costs in terms of what they mean for the daily lives of Canadians, for their energy bills, for the equipment that they buy, for the infrastructure of their economies, for the development and health of their communities, and even for their own incomes and lifestyles.

Book Overview

This book has eight chapters. In Chapter 1 we provide a context for the analysis of GHG emission reduction costs by reviewing the threat of climate change and the international and Canadian policy responses. We first describe the dynamics of the Earth and its atmosphere and explain why alteration of these dynamics by increasing anthropogenic GHG emissions is raising concerns. While adaptation to climate change may eventually be required, mitigation is an option for today that may reduce future risks. International negotiations have resulted in the Kyoto agreement of 1997 in which most developed countries have committed to substantial mitigative efforts between 2000 and 2010. But the cost of mitigation is still poorly understood, even by experts.

Defining the cost of GHG emission reduction is more complicated than it might seem. In Chapter 2 we explain all of the assumptions required in estimating costs. These assumptions include the selection of a reference (or business-as-usual) case, the definition of costs, the forecast of the direction and pace of technological change, the estimate of the simultaneous pursuit of many mitigative actions, the method of determining macroeconomic effects, how costs are allocated among groups in society depending on the choice of mitigation policy, the estimate of other benefits and costs such as changes in local air quality, and the effect on Canada's costs of policies determined by the United States and by further international negotiations. While some of these complications are addressed by testing alternative assumptions, others require key methodological choices.

Cost estimates of GHG mitigation are bedevilled by a methodological split between approaches that focus on emerging technologies and their significant potential for emission reduction, on the one hand, and approaches that focus on human behaviour and its inertia, on the other. The former approaches tend to find that substantial GHG abatement can be inexpensive if we accelerate technological evolution, while the latter tend to find that some new technologies may be costly to press on businesses and consumers, leading to high overall costs of emission reduction. In Chapter 3 we present our methodology for simulating government policies to reduce emissions and for estimating the resulting costs. Ours is a hybrid modelling approach that is technologically explicit yet also incorporates consumer and firm preferences in simulating how such technologies are selected in the marketplace. This approach allows us to probe the potential for cost reduction as new technologies are commercialized and disseminated. It also allows us to test alternative assumptions about consumer and firm receptivity to such technologies in terms of the effect on cost estimates. In Canada's National Climate Change Process, our model was one of several selected to estimate GHG emission reduction costs. We tested actions, policies, and scenarios issuing from this national process and then conducted further tests and sensitivity analyses on our own.

In Chapter 4 we present our national cost estimates. They are disaggregated sectorally in Chapter 5 and regionally in Chapter 6. At the national level, we found that, if a GHG cap and tradable permit system had been applied nationwide in 2000, to achieve the Kyoto target of reducing emissions by 6% from their 1990 levels in 2010, then the resulting permit price would be about \$120/t CO_2e. For Canadian residential consumers, this translates into final energy price increases of 2% to 84% for electricity, 40%-90% for natural gas, and 50% for gasoline. These significant price increases reflect the difficulty of motivating action. However, the present value of total, direct GHG reduction costs is \$45 billion. When these direct costs (investment and energy) are tested in a macroeconomic model, the resulting simulation shows economic output down by 3% in 2010, the equivalent of a one-year recession. We then compare our cost estimates to those of other models and to a similar model applied to the same target in the United States. As expected, our cost estimates fall somewhere between the two contrasting approaches to cost estimation. Nonetheless, our energy prices for consumers and firms are rather high and can be explained by the tight time frame of the Kyoto target, which leaves little time for the normal capital stock turnover that helps to lower the costs of technological change.

Policy design for GHG mitigation must overcome this duality between the high energy prices apparently needed to motivate actions by consumers and firms and the relatively modest macroeconomic impacts of such actions. In Chapter 7 we present policy options on a continuum, with noncompulsory policies (voluntarism, information) at one end and compulsory policies (command and control) at the other. While noncompulsory policies are more easily acceptable to the public and industry, they are unlikely on their own to achieve the Kyoto reductions by 2010. Thus, GHG mitigation policies need to retain some of the desirable attributes of noncompulsory policies while ensuring that substantial reductions occur. One possible avenue is the expanded application of emerging, market-oriented regulations that set aggregate emission, energy form, or technological requirements while allowing participants to trade among themselves in determining how much each contributes to the requirement. We describe salient examples of this new approach.

In Chapter 8, the conclusion, we consider again the major sources of uncertainty for cost estimation and focus on how immediate policy initiatives can help to reduce them.

The Cost of Climate Policy

1
The Climate Change Threat: Why Reduce GHG Emissions?

The prospect of climate change is an enormous environmental challenge. What are the unique qualities of the Earth and its atmosphere that make them so amenable to life as we know it and so potentially fragile and susceptible to changes? What are the sources of the gases said to alter climate, and how will the Earth be affected by climate change? How can we deal with this prospect? As a nation of 30 million people in a world of six billion, to what degree are Canadians responsible for these emissions, and how are we participating to remedy them? What price should we be willing to pay?

This chapter reviews climate change science and the resultant policy response in order to provide the overall context for estimating the cost of GHG emission reductions. First we look at the nature of the Earth and its atmosphere, in particular why changes in the concentration of GHGs in the atmosphere could impact the Earth. Sources and types of GHGs are described, as is the history of the human-induced changes to atmospheric GHG concentrations that have led to the present concerns. Then we turn to the history of international policy responses and negotiations and a review of the initiatives that Canada has undertaken since 1990. Finally, we describe in detail Canada's current process to assess its GHG policy options.

After reading this material, one may agree that human-induced climate change is occurring. However, agreement with this position is not necessary for one to see the value of estimating reduction costs. The world is engaged in a classic exercise of decision making under uncertainty, indeed a chronic uncertainty in that, even for those who are confident that we are changing the climate, many specific impacts are unpredictable. Understanding the cost of action is critical information for such a decision-making endeavour.

What Is Climate Change?

The Earth and Its Atmosphere

The Earth has a livable region, known as the *biosphere,* that is extremely thin relative to its 12,000 km diameter. We have little evidence of any other place that could sustain Earth's life forms. What makes our planet unique?

The Earth's distance from the Sun is critical. A difference of 5% in the Earth-Sun distance would lead to a runaway greenhouse effect or runaway glaciation, consigning the Earth to Venus-like high temperatures or Mars-like constant subzero conditions.[1] The Earth's mean temperature of 15°C seldom varies more than 40° to 60° in either direction, and the Earth has sustained temperatures in this range for much of its history. Two of the Earth's characteristics are primarily responsible for maintaining this stability. First, about 70% of its surface is covered by water, whose mean temperature is about 5°C. This massive body of water moderates the Earth's temperature by acting as a heat sink during daylight hours and a heat source at night and by reducing heat concentrations as its currents distribute heat from warmer to cooler areas. Second, the Earth's gravitational field maintains a thin band of gases, known as the atmosphere, that acts as a thermal blanket regulating both incoming and outgoing solar energy during day and night. The interaction between the Earth's bodies of water and its atmosphere regulates its climate. Alter the characteristics of either and climate conditions can also be expected to change.

Figure 1.1 depicts the interactions between the Earth, its atmosphere, and solar radiation. The atmosphere, a layer of gas about 150 km thick, permits most forms of solar electromagnetic radiation to reach the Earth's surface in varying amounts.[2] Upon striking the Earth, the radiation, including its shorter waves known as visible light, is generally converted to heat and

Figure 1.1

Greenhouse effect

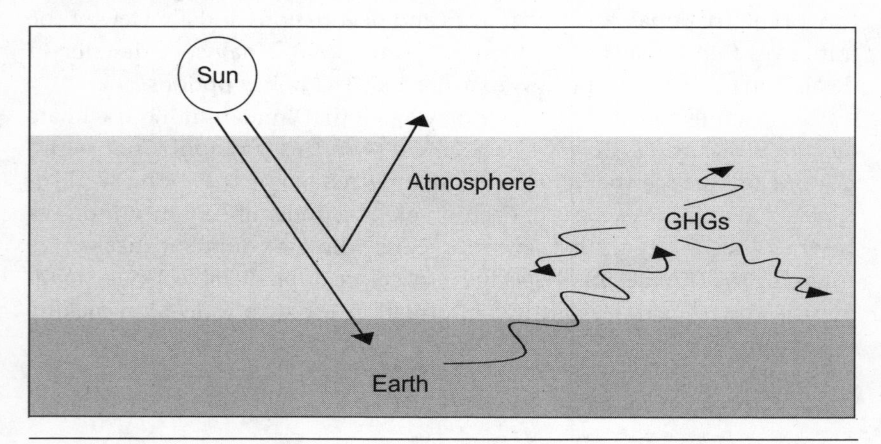

Note: The solid lines represent solar radiation; the curved lines represent infrared radiation.

re-emitted in the form of infrared radiation. The atmosphere does not let infrared radiation leave as readily as visible light radiation enters, so heat energy is retained near the Earth's surface. This *greenhouse effect* maintains the Earth's temperature about 33°C higher than it otherwise would be.[3]

The greenhouse effect is a result of the atmosphere's chemical constituents. Composed primarily of nitrogen and oxygen, the atmosphere also contains trace gases: argon, neon, helium, krypton, xenon, ozone (O_3), hydrogen, carbon dioxide (CO_2), nitrous oxide (N_2O), methane (CH_4), and water vapour, a gas whose local proportion varies considerably from day to day. Table 1.1 defines these constituents and shows how concentrations have changed over time. While their proportion of the Earth's atmosphere may be small, the trace gases play important roles in maintaining the Earth as a livable region.[4] A number of these gases – CO_2, CH_4, N_2O, and a few other gases generated by human activity – trap infrared radiation (heat) and then re-emit it as infrared energy in all directions.

Sources and Types of GHGs, Natural and Human-Induced

The surface of the Earth and its atmosphere are constantly changing. A number of geological and chemical cycles ensure the supply and replenishment of many of the elements and compounds needed to sustain life: key cycles are water, nitrogen, and carbon. Our interest is primarily in the carbon cycle (Figure 1.2), which explains the flow of carbon in its gaseous form, in association with oxygen as CO_2, to fixed organic material known as biomass. When carbon is fixed, it is said to be *sequestered*. The medium in which it is captured and fixed is generally referred to as a *sink*. Thus, one may refer to the forest sink or the ocean sink. The arrows in the diagram show the transfers of carbon between different forms. Most carbon in biomass gets released back to the atmosphere again as CO_2. However, over geological time, the fixation of carbon has exceeded its release as CO_2 because some organic material gets trapped underground in sedimentary strata and, after long periods of sustained pressure, is transformed into fossil fuels – coal, oil, and natural gas. Certain forms of sea life also biologically fix bicarbonate with calcium, ultimately producing limestone and dolomite.

Carbon is returned to the atmosphere in different ways. Oxidative combustion of any carbonaceous compound (wood, oil, methane, coal) generates CO_2 as a by-product. Respiration, the process of obtaining energy from the oxidation of sugars in living systems, generates CO_2 as well. Living systems release CO_2, whether bacteria decomposing organic material, yeast generating alcohol from sugar, or humans metabolizing their lunch. Other noncombustive sources include deforestation and land use change, volcanoes and similar geological activities, and certain chemical reactions such as the calcination of limestone to make cement or lime.

Table 1.1

Atmospheric gases, global warming potentials, and sources

Gases	Concentration, 1992[a]	Concentration, pre-industrial[a]	CO_2e rating	Lifetime in atmosphere, remarks
Nitrogen (N_2)	78.084%			
Oxygen (0_2)	20.946%			
Argon (Ar)	0.934%			
Neon (Ne)	0.182%, 182 ppmv			
Helium (He)	0.0524%, 52 ppmv			
Water vapour (H_2O)[b]	0 to 2%		–	
Carbon dioxide (CO_2)[b]	0.35%, 350 ppmv	0.28%, 280 ppmv	1	Variable, normally 200 years, increase almost entirely of human origin
Ozone (O_3)				Not emitted directly by humans but influenced by human activity. Increased tropospheric O_3 concentrations contribute to warming, while stratospheric O_3 depletion has a net cooling effect.
In troposphere[b]	0.02 to 0.1 ppmv		–	
In stratosphere[b]	0.1 to 10 ppmv		–	
Methane (CH_4)[b]	1.74 ppmv	0.70 ppmv	21	12.2 years, natural and human origin
Nitrous oxide (N_2O)[b]	0.311 ppmv	0.275 ppmv	310	120 years, natural and human origin
CFC-12 (CF_2Cl_2)[b]	0.503 ppbv	0.0 ppbv	8,500	102 years, entirely human origin
CFC-11 ($CFCl_3$)[b]	0.3 ppbv	0.0 ppbv	4,000	Entirely human origin
Sulphur hexafluoride (SF_6)[b]	0.032 ppbv	0.0 ppbv	23,900	3,200 years, entirely human origin
Halon-1301 ($CBrF_3$)[b]	2.0 pptv	0.0 pptv	5,600	Entirely human origin
Halon-1211 ($CBrClF_2$)[b]	1.7 pptv	0.0 pptv	–	Entirely human origin
Nitric oxide (NO)				
In troposphere	0 to 1 ppbv			
In stratosphere	Up to 0.02 ppmv			

a The concentrations of atmospheric gases are given in either percentage by volume (which is the same as parts per hundred by volume), parts per million by volume (ppmv), parts per billion by volume (ppbv), or parts per trillion by volume (pptv).

b These gases have GHG properties of heat retention. Their global warming potentials are indicated by a CO_2 equivalency (CO_2e) rating as of 1995, except for ozone and water vapour, whose radiative role is complex and less understood.

Source: Statistics Canada, *Human Activity and the Environment 2000* (Ottawa: Statistics Canada, 2000).

Figure 1.2

The carbon cycle

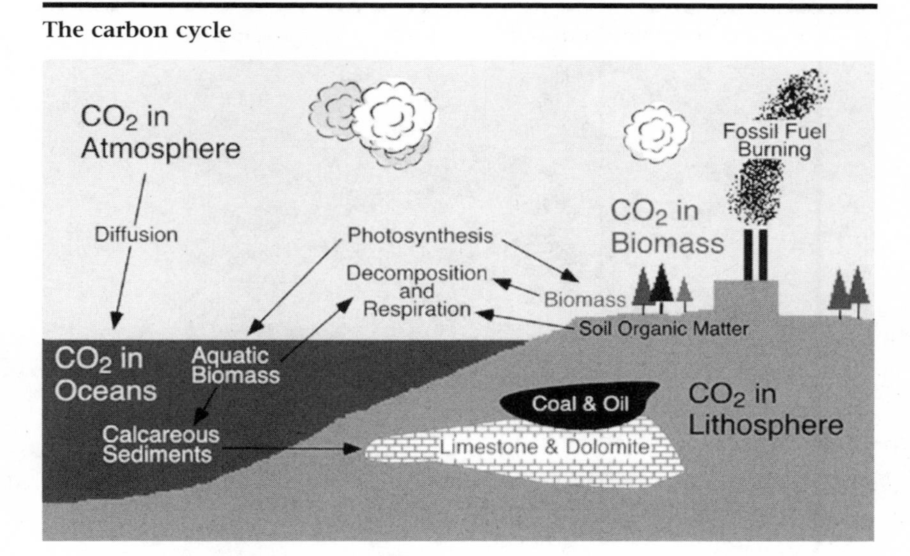

Source: M. Pidwirny, Department of Geography, Okanagan University College, Figure 9r-1, *Fundamentals of Physical Geography* (1996-2000), <www.geog.ouc.bc.ca/physgeog>.

The carbon cycle is generally found to be in equilibrium, and it dwarfs the changes caused by human-induced, or *anthropogenic,* carbon releases to the atmosphere. Indeed, some skeptics point to the relatively small anthropogenic contribution as evidence that humans are not causing climate change. However, according to climate specialists, small changes in atmospheric concentrations of GHGs, carbon dioxide in particular, can have significant impacts on the atmosphere's greenhouse function. But just how significant is an uncertainty likely to remain for some time.

Some skeptics suggest that we need not take action to reduce our emissions because natural sinks will absorb the extra carbon in the atmosphere. Thus, sinks can act like buffers or negative feedback loops if their rate of CO_2 uptake increases as the atmosphere becomes more CO_2 enriched. This might especially be the case for oceans, the largest CO_2 sink, which are teeming with microscopic organisms and hold vast amounts of CO_2 in solution. However, many scientists working in this area believe that only about half of anthropogenic CO_2 is absorbed in this way, while the rest remains in the atmosphere – which explains the steady increase of atmospheric CO_2 concentrations.

It is not only CO_2 that acts as a thermal blanket. A number of other GHGs are even more efficient at retaining heat. They include common gases such as methane (CH_4) nitrous oxide (N_2O) and a long list of less common, human-produced compounds such as halocarbons: compounds of carbon

Figure 1.3

Contribution of human-induced gases to the greenhouse effect

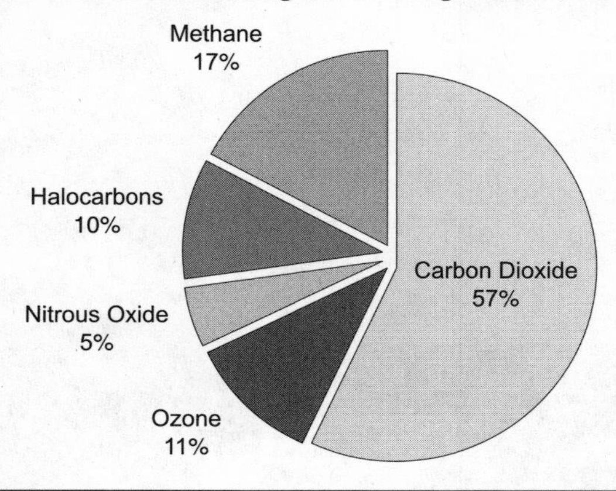

Sources: Calculations based on information in Intergovernmental Panel on Climate Change, *Climate Change 1995: The Science of Climate Change* (Cambridge: Cambridge University Press, 1996); Pembina Institute, "Greenhouse Gases," <www.climatechangesolutions.com/english/science/gases.htm>.

with fluorine, chlorine, or both. Each compound has been analyzed in terms of its ability to retain heat (*radiative forcing*) compared with CO_2; this is described as a *CO_2 equivalency (CO_2e) rating* or *global warming potential*. For example, CH_4 has a CO_2e rating of 21, meaning that one tonne of CH_4 has the same global warming potential as 21 tonnes of CO_2. The estimations are based not only on ability to retain heat but also on life span in the atmosphere and other characteristics. Table 1.1 includes CO_2e ratings for different GHGs.[5]

Although there are only small amounts of GHGs in the Earth's atmosphere, they trap a significant part of the heat that is emitted as radiation by the Earth's surface. Water vapour is the most abundant and important GHG and is responsible for about 60% of the total greenhouse effect. The remaining 40% is due to the gases discussed above that are closely linked with human activity: CO_2, CH_4, N_2O, halocarbons, and ozone. Their respective contributions to the human-induced greenhouse effect are shown in Figure 1.3. Although ozone is better known for the role that it plays in the ozone layer, human-induced changes in the concentration of tropospheric (ground-level) ozone relative to stratospheric ozone may be strengthening the greenhouse effect.

History of GHGs and Human Activity

Prior to the industrial revolution, people depended primarily on renewable sources of energy: muscle power (human or domesticated animal), flowing water, wind, and the combustion of plant and animal wastes. But with the development of the steam engine at the birth of the industrial revolution, the use of coal and eventually other fossil fuels contributed to profound changes in production processes, farming, and domestic activities. Since 1750, and especially during the past 100 years, the use of fossil fuels has increased dramatically. Today fossil fuels provide 83% of the energy used by member countries of the Organization for Economic Cooperation and Development, releasing roughly 28 billion tonnes of CO_2 into the atmosphere in 2000. Ice core samples that go back 400,000 years indicate that levels of CO_2 are at a peak today and that the rate of increase in the Earth's atmospheric concentration is the fastest ever. In fact, the increase over the past 150 years equals that which occurred from the height of the last ice age 20,000 years ago to the year 1750.

Researchers using models to predict future atmospheric concentrations of CO_2 estimate that by the year 2100 the CO_2 content of the atmosphere will have doubled or tripled over pre-industrial levels. The margin of error around these values is estimated to range from 10% to 30% because of the high uncertainty about the buffering effect of sinks.

Fossil fuel use as well as industrial and agricultural activities generate other GHGs. The atmospheric concentration of methane has increased 151% since 1750 and the concentration of nitrous oxide by 17% to its highest level in at least the past 1,000 years. While CO_2 emissions are closely linked to the carbon in fuels, these other GHG emissions depend on a variety of factors such as industrial process or combustion technology. Nitrogen, by far the largest component of the atmosphere, combines with oxygen quite readily under heat and pressure, typical of internal combustion engines and some boiler systems. Industrial processes also release chlorofluorocarbons (CFCs), hydro-chlorofluorocarbons (HCFCs), perfluorocarbons (PFC), and sulphur hexafluoride (SF_6), each a potent GHG. Many of these gases remain in the atmosphere for a long time, giving them CO_2e indices in the thousands (Table 1.1).

Possible Impacts of a Change in GHG Concentration in the Atmosphere

At first glance it appears unlikely that such minor changes in the composition of the Earth's atmosphere (fractions of a percent) would have significant effects. However, global temperature estimates over the past five centuries indicate a rising trend (Figure 1.4), and it is estimated that temperatures have increased by 0.6°C over the past century. In its 1995 report, the Intergovernmental Panel on Climate Change (IPCC) stated that "the balance of evidence suggests a discernable human influence on global climate," and it strengthened that statement in its 2000 report by stating that "there is now new and stronger evidence that most of the warming observed

Figure 1.4

Trend in the Earth's average annual temperature

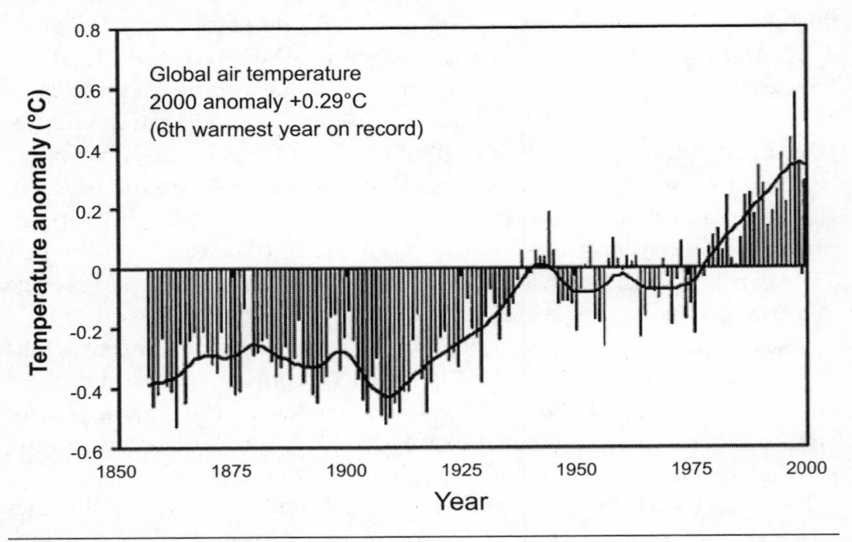

Note: Further details about the development of the data set are available in P.D. Jones et al., "Surface Air Temperature and Its Changes over the Past 150 Years," *Review of Geophysics* 37 (1999): 173-99.
Source: Climatic Research Unit, University of East Anglia, <www.cru.uea.ac.uk/cru/info/warming/>.

over the last 50 years is attributable to human activities."[6] The IPCC goes on to forecast a further increase in average global temperatures of 2°C by 2100, although with uncertainties this increase could range from 1°C to 4°C.

What would be the effects of such temperature changes? An increase of 2°C appears small to a Canadian, who might experience temperatures that vary by 60°C through the seasons of a typical year. However, it is estimated that at the depth of the last ice age average global temperatures were only 3.5°C to 5°C colder than at present. Many climate and ecology experts believe that increases in the average global temperature of 1°C to 4°C in the short span of a century could have profound impacts on climate and ultimately ecosystems.

Climate change is likely to impact the hydrological cycle, thus affecting the patterns and intensities of precipitation and drought, the conditions for agriculture and forestry, freshwater availability, and overall ecosystem viability. There is a potential for coastal flooding and even submerging of island states as sea levels rise. More subtly, changed moisture regimes could expand forested regions and transform semi-arid regions into deserts, causing the extinction of species and increasing the abundance of others, including the expansion of disease-causing organisms into new areas.[7] Of particular interest to Canadians is that permafrost zones are predicted to move steadily northward and detrimentally alter the habitat and perhaps the survival of caribou and other inhabitants of the tundra. Additionally, warming oceans could negatively impact salmon populations while extending the range of species such as mackerel.

The rate of such change could be many times greater than the rate at which species can adapt. Higher global temperatures may also increase the likelihood of extreme climatic events. While evidence is not conclusive, some climatologists consider it likely that the world will see increased intense precipitation events in the northern hemisphere and increased cyclone activity in tropical regions.

The impact on humans is controversial and not well understood. Many of these events could result in considerable economic burden for a nation, region, or local population. Others could result in significant benefits. For example, the impact on food crops is not well understood, and estimates vary greatly; some studies suggest declines of 80% or more, while others forecast increases of over 200%.[8] Flooding and higher sea levels could force internal or international migrations of human populations. Human health could be affected because of harsh changes in the local climate (heat waves, levels of precipitation, drought, extreme weather events) and ecological disturbances that include the spread of tropical diseases, parasites, and crop pests.

Dealing with Climate Change

Governments are increasingly convinced by those who argue that we must take seriously the threat of climate change. Many have decided that the risk of negative impacts is real and that we must act now to reduce the risk; the perspective is changing from one of "Should we act?" to one of "How should we act?" In essence there are two ways of acting: either to adapt to the changes or to mitigate them.

Adaptation

Adaptation is a response to change in the environment in an effort to minimize adverse effects. The ability to adapt depends on the extent to which climate change can disrupt the system and the extent to which the system can be modified to avoid damage from disruption.

Adaptation could include more efficient management of resources; construction of dikes; improved monitoring and forecasting systems for floods and droughts; construction of new reservoirs to capture and store excess flows produced by altered patterns of snowmelt and storms; changes of individual crops and crop mixes; improved water management and irrigation systems; and changes in planting schedules and tillage practices. The extent of adaptation will depend on the affordability of such measures, access to know-how and technology, the rate of climate change, and bio-physical constraints such as water availability, soil characteristics, and crop genetics.

Mitigation

Mitigation is an effort to maintain the existing or historical state by stopping or counteracting the source of a change. Examples are to stop combusting fossil fuels and to sequester CO_2 emissions from such combustion.

While it is not clear what an acceptable level of GHGs in the atmosphere might be, atmospheric scientists claim that simply to stabilize the atmospheric concentration of CO_2 would require a decrease of at least 70% in the release of anthropogenic emissions, an achievement that clearly would take decades at best.[9] Because of time lags in the absorption processes of sinks, the greenhouse effect would intensify during this period and for some time after the targeted reduction has been achieved. International agreements that stabilize global anthropogenic emissions (as opposed to atmospheric concentrations) can only slow climate change because current emissions are already at levels that cause rapid increases in atmospheric concentrations. The Kyoto agreement would have even less effect because it stabilizes only the emissions of developed countries and a few others, so this stability would be offset by rising emissions from developing countries, China and India in particular.

In developing a mitigation strategy, it is important to specify the key factors that determine the level of emissions. A useful tool is a *decomposition equation* that explains changes in GHG emissions in terms of changes in specific factors. A common term for this in the GHG literature is the *Kaya Identity,* shown in Figure 1.5. Each term provides insight into how one might reduce emissions.

The first term relates GHGs to energy use (GHG/E). A change in this term may occur if the share of fossil fuels has declined in the energy mix, say through the greater use of renewables or nuclear power. It may change if there is a shift among fossil fuels such as from coal toward natural gas. It may also change if technologies are used to separate CO_2 emissions from fossil fuels (either before or after combustion) and prevent them from entering

Figure 1.5

Kaya Identity decomposition equation

$$\%\Delta GHG = \%\Delta\frac{GHG}{E} + \%\Delta\frac{E}{Q} + \%\Delta\frac{Q}{P} + \%\Delta P$$

Note: GHG = greenhouse gas, *E* = energy, Q = output of the economy, and *P* = population.

the atmosphere. Carbon can be captured and sequestered by injection into exhausted oil and gas reservoirs, active wells to enhance oil and gas recovery, coal beds to release methane, or deep saline aquifers. Forestry and agricultural carbon sinks can also be enhanced through forest management, cropland management, and revegetation actions.

The energy intensity term (E/Q) reflects changes in energy used per unit of economic output. If the energy efficiency of equipment, buildings, and infrastructure changes, then it will show up in this term. However, other factors – not specifically related to technical energy efficiency – can also cause this ratio to change, which is why the preferred term is *energy intensity* instead of energy efficiency. The measure of output (Q) may change even though it has not changed in physical terms. Output is usually measured in monetary units, corrected for inflation, in order to estimate changes in physical output. However, if the value of the sector's output has changed at a different rate than inflation, then this term can change even when the physical product does not change.[10] When it comes to personal energy uses, such as home heating or vehicle travel, even lifestyle changes can show up in this term. A deliberate lowering of home heating temperatures, or reduced vehicle use for short, less efficient urban trips, can lead to changes in energy intensity. Because some sectors are more energy intensive than others, structural shifts in the economy will also change energy intensity and thus GHG emissions. For example, a shift toward a more service-oriented economy would reduce energy intensity and GHG emissions.[11]

The economic output per capita term (Q/P), referred to as income or standard of living, is conventionally measured as the monetary value of economic output divided by population. Generally, one assumes that greater income is associated with greater physical output, greater energy use, and thus greater GHG emissions. However, increases in Q/P may not lead to proportionately higher GHG emissions if the higher incomes are linked to advanced, energy-efficient technologies and a structural change toward a services and information economy, both of which reduce E/Q. In other words, the terms in the Kaya Identity are not necessarily independent of each other.

Finally, if all other terms in the identity are held constant, then increases in population will lead to corresponding increases in GHG emissions.

We have been describing thus far an energy-focused decomposition equation. But GHG emissions are not just associated with energy use. One could add functions to isolate emissions related to other GHG-producing activities. For example, there could be a similar identity for forestry or agricultural GHG emissions and sinks and yet another for solid-waste-related GHG emissions.

Mitigation or Adaptation?

Adaptation and mitigation approaches are not mutually exclusive; we may

Table 1.2

Time scales of processes that influence or are influenced by the climate system

Process	Time scale
Turnover of capital stock	Years to decades
Stabilization of long-lived GHGs	Decades to millennia
Equilibration of the climate system	Decades to centuries
Equilibration of sea levels	Centuries
Restoration/rehabilitation of disturbed ecosystems[a]	Decades to centuries

a Some changes such as species extinction are not reversible.
Note: These estimates assume stabilization of GHG emissions.
Source: Intergovernmental Panel on Climate Change, *IPCC Second Assessment: Climate Change 1995* (Cambridge: Cambridge University Press, 1996), 28.

require both adaptive and mitigative strategies given that we have already changed the atmosphere significantly. Indeed, adaptive approaches are already under way, as witnessed by the growing focus on climate change factors by insurance companies in their claims forecasting and premiums setting.

One reason that adaptation must be considered seriously is that GHGs can continue to have an effect for many years after entering the atmosphere. First, any increased uptake of GHGs by the biosphere from the GHG-enriched atmosphere can take a long time depending on a variety of factors. Second, the deep oceans are slow to adjust to temperature change simply because of their large mass.[12] Third, the response of ice sheets to changes in the average global temperature is also slow. These processes will be responding to previous and upcoming temperature changes long after humankind has managed to reduce emissions or even to stabilize atmospheric concentrations.

Table 1.2 provides estimates of the time scale of the processes that influence or are influenced by the climate system. Given this inertia, it is likely that climate policies must be a combination of adaptation and mitigation. Adaptation strategies are more a challenge for the future as the more significant impacts of climate change emerge. Mitigation, in contrast, requires action now if it is to have the effect of reducing some of the more extreme risks associated with climate change. This is the motivation for current research on the costs of mitigation and for international and national negotiations to reduce emissions. It is the rationale for this book.

Responsibilities for Human GHG Emissions: The World and Canada

Action at an international level to mitigate climate change has involved significant wrangling over the responsibility for and the allotment of emission reductions among the world's nations. Countries can be ranked differently

Table 1.3

Cumulative and average CO_2 emissions of major emitters

Region	Cumulative CO_2 emissions (1950 to 1995 Gt)		1995 CO_2 emissions (Gt)		CO_2 emissions per capita (1995 t/capita)	
World	662.24		22.32		4.62	
United States	180.24	1	5.16	1	19.4	1
European Union	125.22	2	2.98	3	9	7
Russia	66.69	3	1.82	4	12.1	3
China	54.03	4	3.19	2	2.6	10
Germany	41.78	5	0.84	7	10.3	4
Japan	29.74	6	1.13	5	9.2	5
United Kingdom	26.67	7	0.54	8	9.2	5
Ukraine	20.93	8	0.44	9	8.4	8
France	16.44	9	0.34	11	5.9	9
India	14.51	10	0.91	6	1.1	11
Canada	14.47	11	0.44	10	14.7	2

Source: E. Claussen and L. McNeilly, *Equity and Global Climate Change: The Complex Elements of Global Fairness* (Arlington, VA: Pew Center on Global Climate Change, 1998), Appendix.

according to whether one considers their cumulative CO_2 emissions over a common historical period, their share of current global CO_2 emissions, or their CO_2 emissions on a per capita basis. Table 1.3 ranks major emitters according to these categories. Canada ranks 11th in terms of cumulative emissions and 10th when annual emissions in 1995 are considered. Although Canada is ranked second in per capita CO_2 emissions among the major emitters in this table, its ranking is 10th when all countries of the world are included (ranking behind countries such as Australia, Singapore, Norway, and Luxembourg).

There are several reasons for Canada's high per capita emission levels. Not only is Canada a world supplier of primary commodities, but also it has a small population living in a large country with a cold climate. Producing goods, travelling great distances, and keeping warm consume a lot of energy. Canada has significant, high-quality energy resources, including substantial reserves of coal, oil, and gas, primarily in the western region.[13] With energy supply plentiful and relatively inexpensive, Canada has an energy-intensive economy and lifestyle even when compared with other countries with similar geographic and demographic characteristics. For example, Canada uses significantly more energy per capita than equally cold and thinly populated Scandinavian countries such as Sweden and Finland.

One must consider a number of factors in determining how responsibility for emission reduction is allocated.

Figure 1.6

Forecast of CO_2 emissions to 2015 in China, the United States, and Canada

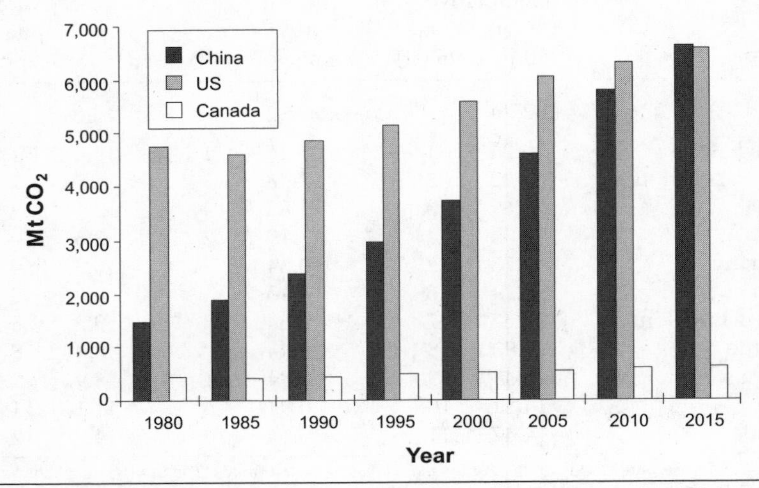

Source: US Energy Information Administration, *China: An Energy Sector Overview,* 1997, <www.eia.doe.gov>.

Historical sources: Developing countries point out that in most cases they have not contributed to the problem as much as the industrialized world has, and one of the means to become developed would be denied them if fossil fuel use were restricted. Thus, it would be inequitable to saddle developing countries with the same criteria and restrictions that one would apply to developed nations.

Future sources: If present trends continue, where will GHGs be produced in the future? CO_2 emission growth appears to be dropping in developed countries and rising in developing countries. According to recent US Department of Energy studies, China will surpass the United States in CO_2 generation by 2015 to become the single largest contributor of CO_2 emissions in the world. Figure 1.6 provides a forecast of emissions.

Geography and geology: What roles do climate and resource characteristics play? Colder climates obviously require greater use of energy for space heating. However, as developing countries in hot climates become wealthier, they will experience a dramatic increase in energy use for air conditioning, just as this is an important energy service today in the southern United States. In terms of resources, countries such as Canada and Australia, with small populations, exploit significant quantities of the world's raw materials. Extraction and upgrading of these resources for export as manufacturing inputs (aluminum, pulp, metals) tend to be energy intensive. In contrast, Japan has few natural resources, has a large population, and is engaged in processing these inputs into finished manufactured goods, an activity that is much less energy intensive. This contrast is significant in explaining country differences in emission intensities. Another factor is whether fossil-fuel-rich countries are endowed with coal, oil, or natural gas. China, for example, has immense reserves of coal (12% of total world reserves) but limited supplies of natural gas.[14] Coal-using technologies today emit more CO_2.

International Response to the Climate Change Threat

In February 1979 the first World Climate Conference was held in Geneva. Organized by the World Meteorological Organization, it focused on how climate change might impact human activities. The conference led to the establishment of the World Climate Program. In 1985 a group of concerned scientists from 29 developed and developing countries met in Villach, Austria, on the initiative of the United Nations Environment Programme, the World Meteorological Organization, and the International Council of Scientific Unions, to discuss whether or not human activity was affecting the Earth's climate. They concluded that the evidence pointed to a significant probability that human activity was indeed changing the Earth's climate.

This was followed by a series of conferences and workshops attended by politicians as well as scientists and international organizations. At a conference in Toronto in 1988, attendees established a target to reduce global CO_2 emissions by 20% of 1988 levels by 2005. The same year, the UN General Assembly approved the establishment of the Intergovernmental Panel on Climate Change (IPCC). The IPCC produced its first assessment report in 1990, concluding that the possibility of global warming had to be taken seriously.

In 1992 at the Rio de Janeiro Earth Summit, the United Nations Framework Convention on Climate Change was signed by 165 states, including the United States. It set a voluntary goal to cut CO_2 emissions to the 1990 level by the year 2000, with the ultimate objective of achieving stabilization of GHG concentrations at a level that would prevent dangerous anthropogenic interference with the climate system. Article 4 of the convention stated that both developed and developing countries should adopt mitigation strategies and promote further analyses of global warming and climate change, although industrialized countries should take the lead. In 1995 in Berlin, the first Conference of Parties decided to relieve developing countries of obligatory actions. The IPCC's reports in 1995 and 2001 supported the position that humans are influencing the climate, and mitigation action should therefore be undertaken.

The Kyoto Protocol

In December 1997 over 160 countries met in Kyoto, Japan, for the third Conference of Parties, and 84 of those countries negotiated an agreement

Table 1.4

Emissions limitation or reduction commitment of different regions under the Kyoto Protocol (1990 emissions = 100%)

Region	Commitment (%)	Region	Commitment (%)
Australia	108	New Zealand	100
Canada	94	Norway	101
European Union	92	Poland	94
Hungary	94	Russian Federation	100
Iceland	110	Ukraine	100
Japan	94	United States	93

Source: United Nations, *Kyoto Protocol to the United Nations Framework Convention on Climate Change,* Conference of the Parties, Third Session, 1998, Annex B.

that would reduce their aggregate GHG emissions to 5.2% below 1990 levels by the period 2008-12. Annex I countries were allotted different targets (Table 1.4).[15] The European Union received a target of 8% below 1990 emissions, the United States 7% below, Canada 6% below, and Australia 8% above.[16] International agreements made by negotiators require domestic ratification. It was agreed at Kyoto that the protocol would not come into effect until ratified by 55 countries that collectively account for 55% of Annex I emissions.

The protocol specifies a number of mechanisms through which the targets may be attained.

- *Domestic emission reduction:* This is action to reduce GHG emissions undertaken within a country's borders. Such actions tend to be fuel switching and energy efficiency but may include CO_2 emission capture and sequestration.
- *Augmentation of domestic sinks:* A country may augment the carbon storage potential of its natural sinks by changing management practices on forested and agricultural land.
- *Joint implementation:* This is a method of reducing GHG emissions in one country with the help of another country. The benefits of the reduction are shared between the two countries, both of which have emission obligations.
- *Clean development mechanism:* This is a modified version of joint implementation that gives credit to Kyoto signatories for undertaking GHG emission reduction projects in developing countries that are not signatories.
- *Emissions trading:* Signatory countries can develop GHG emissions-trading mechanisms that would serve not just for domestic implementation but also, ultimately, as a means for the cost-effective, international coordination of efforts to reduce GHG emissions. In this sense joint implementation

and the clean development mechanism are seen as specific prototypes to be replaced one day by a more general trading mechanism.

Since the signing of the Kyoto Protocol, the signatories have continued to meet to discuss the many complexities of constructing and ratifying an international agreement of this scope and to more precisely define the roles of the above mechanisms. Certain issues have received considerable attention and controversy.

The Kyoto Protocol gives countries such as Russia and Ukraine, which have closed many high-emission factories since 1990 in the transition to a market economy, pollution quotas that permit them emissions in excess of current levels. Some countries would like to buy these quotas and thus lower their domestic emission reductions while still meeting their targets. But some observers fear that cash-starved countries will sell their pollution quotas and then generate emissions anyway as their economies grow. They argue that safeguards are needed and that the total amount of reduction credit from this practice should be limited if meaningful global reductions are to be achieved.

The definition and the role of sinks have been controversial. What actually is a sink? Could one include the extra carbon stored in soils consequent to altered farming and forestry practices? How would one measure and credit such a change? The resolution of these issues has been important to countries, such as Canada, Australia, and the United States, that have large land areas and extensive potentials for greater absorption of carbon with altered land management. A broad definition of sinks greatly reduces the amount of emissions reductions needed from energy production and use.

A related issue is the role of flexibility mechanisms to meet national targets. Members of the European Union have argued that most reductions by wealthy countries should come from domestic actions, while the United States, Japan, Australia, and Canada have pushed for greater flexibility. The focus on domestic GHG reductions in Europe can be interpreted as coinciding conveniently with European countries' move away from domestic coal use since 1990. In Great Britain coal use has declined dramatically, particularly after much of the coal-mining industry and many coal-fired power plants were closed in the years following electricity market reform in 1989. Likewise, Germany has experienced a windfall reduction as the country's unification in 1990 led to the closure of inefficient industrial and coal-using electricity plants in the former East Germany. In response, the Europeans argue that, because their economies in 1990 were already much less carbon intensive than those of the United States and Canada, they face substantially higher incremental costs for making additional reductions.

Debate has also occurred about how the Kyoto Protocol will be enforced once it is ratified. What would occur if a nation does not attain its specified target? How does one induce compliance and punish noncompliance? The Europeans have promoted the use of trading constraints and other economic sanctions against noncompliant nations, while the United States has favoured a penalty of additional obligations in subsequent commitment periods.

The Bush administration's intention not to ratify the Kyoto Protocol is based on its concern with the overall scope of the agreement rather than with the specific implementation issues described above. More specifically, the United States argues that the Kyoto Protocol will incur high costs for a few countries like the United States without making much headway on global emissions. Without the inclusion of the entire global community, emission reductions by developed countries will be quickly offset by rapidly growing emissions from developing countries. The United States also points to significant competitive issues. How can a country burdened with the costs of reducing GHG emissions compete fairly with one having no such restriction? What would prevent industry from relocating to regions where GHG reduction is not required, a phenomenon known as leakage?[17] The United States has instead sought the reopening of international negotiations, which it hopes will lead to a more comprehensive global commitment.

Other Kyoto signatories have not turned against the protocol, but negotiations about implementation and verification continue. The Conference of Parties meeting in Bonn in the summer of 2001 reached agreement on many key issues, although important technical details remain to be resolved.

A critical concern for Canada and some other signatories was the ruling on how much of a country's emission reduction commitment could be achieved by nondomestic (supplemental) actions. At Bonn, parties agreed that with the use of flexibility mechanisms countries could meet a significant share of their commitments, although the exact amount was not specified. The Canadian government has interpreted this to mean that it should attain a minimum of 50% of its target domestically, which means that it might purchase the other 50% from other signatories via flexibility mechanisms. This interpretation may change.

Most activities claimed as sinks will be eligible for credits, as long as they have occurred after 1990. This eligibility includes forest management, cropland management, and revegetation. There is no overall cap on sink credits, though specific limits have been established for forest management.

Although progress has been made in addressing many outstanding matters, much remains to be done. The specific issues raised by the Kyoto Protocol are symptomatic of broader concerns facing any effort to achieve a global response to the climate change threat.[18] These concerns will need to be addressed in order to advance a multilateral international response and to negotiate more substantial reductions in the future.[19] Any progress in this direction must ultimately come to terms with two competing concerns:

recognition of the perils of human experimentation with the Earth's atmosphere and reluctance to take actions that may cause significant financial and lifestyle costs for the present generation without certainty of future benefits. Analyses such as the one presented in this book can illuminate the trade-offs between these two concerns and assist countries such as Canada in determining their positions as the world continues to grapple with this complex challenge.

Mitigation Initiatives in Canada

A broad spectrum of views characterizes Canadian reaction to the climate change threat. Not only does Canada face the challenging task of achieving cooperation between divergent interests, but also it must do so within the framework of Canadian federalism. Under the Constitution, the environment is a shared federal-provincial jurisdiction: the management of natural resources lies within the jurisdiction of the provincial governments, while international agreements are a federal responsibility. This division of powers makes it difficult for Canada to move forward with a major mitigation initiative.[20]

Canada's national effort is centred on the National Climate Change Process (NCCP) initiated in early 1998 at a meeting of federal, provincial, and territorial ministers of energy and the environment. It is a forum that provides overall leadership and guidance on Canada's response to the climate change issue.[21] In October 2000 all governments except that of Ontario approved a National Implementation Strategy on Climate Change based on the work of the NCCP. This strategy is composed of three phases, delineated by various degrees of uncertainty.[22] Phase 1, Action under Uncertainty, represents the period until Canada ratifies the Kyoto Protocol or some subsequent agreement. This phase contains cost-effective measures that deliver important ancillary health, economic, and environmental benefits and lay the groundwork for further action. Phase 2 defines additional measures that Canada should undertake, given the results of additional analyses, and Phase 3 will begin if Canada ratifies the Kyoto Protocol or some substitute commitment.

Many of the federal government's departments participate in and support the NCCP. Two have been allocated primary responsibilities: Natural Resources Canada is in charge of GHG evaluation and response, and Environment Canada is responsible for developing Canada's GHG inventory. As a signatory to the Kyoto Protocol, Canada must submit GHG inventories to the United Nations Framework Convention on Climate Change and establish a history of CO_2 and other GHG emissions back to 1990. Other federal agencies participate by reviewing climate change impacts and mitigation options in their particular fields. Thus, Agriculture Canada plays a major role in the assessment of soil sinks and the impact of agricultural practices on GHG capture and emissions. In a similar manner, the Canadian Forestry

Service has been assessing the relationship between forests and carbon sinks in soils and forestry products. Transport Canada participates in the assessment of transportation's impact on climate change.

Regional response to climate change varies. As of October 2001, three provinces and one territory – British Columbia, Ontario, Quebec, and the Northwest Territories – have produced climate change action or business plans. These plans outline specific actions taken during the first phase of the National Implementation Strategy. In addition, a number of provincial governments, in conjunction with two federal ministries, have developed a pilot project on emissions trading. Called the Greenhouse Gas Emission Reduction Trading Pilot, it started in June 1998 to provide experience with a market-based approach to emissions reduction.[23]

In general, voluntary approaches have dominated Canada's response to climate change. One of the government's first major initiatives, the Voluntary Challenge and Registry, encourages private and public sector organizations to voluntarily limit their net GHGs, not only as good citizens and for private economic benefit but also to avoid or at least delay more stringent government intervention. In this program, firms are asked to submit an action plan that contains senior management support, a commitment to regular reporting, and a base year calculation. The action plan can be much more detailed and provide specific activities to be undertaken in the future. Each company submits progress reports on its action plan. Table 1.5 displays the growth of the registry over time. The companies that submit are responsible for roughly 75% of all industrial GHG emissions. Registrants include about 75% of all pulp and paper firms, 95% of all cement producers, and 100% of all petroleum refineries.

A number of industry associations generate status reports that define the activities and progress made by their members. Each year, for example, the

Table 1.5

Progress report, industry submissions to Voluntary Challenge and Registry, Inc.

Year	Action plans	Progress reports
1995	94	4
1996	331	88
1997	354	112
1998	547	168
1999	681	258
2000	757	344

Source: Online table, see Voluntary Challenge and Registry, Inc., Website, <www.vcr-mvr.ca>.

Canadian Chemical Producers' Association publishes its *Responsible Care* report and its *Reducing Emissions 8* report, part of the National Emissions Reduction Masterplan under its Responsible Care program. These annual documents review member activities and track all emissions, including GHGs. The Canadian Petroleum Products Institute also publishes an annual *Environment and Safety Performance Report* that displays, among other things, its efforts to reduce GHG emissions since 1990.[24]

While participation in the Voluntary Challenge Registry may be high, its effectiveness is not universally accepted. In a review of the registry, the Pembina Institute, a nongovernmental environmental organization, determined that it was inadequate in terms of reaching Canada's Kyoto Protocol target and showed no sign of improving prospects.[25] The institute recommended that the reporting of GHG emissions by significant industrial emitters be mandatory in Canada and that complementary financial incentives and regulatory instruments be invoked to induce greater response.

The National Climate Change Process

Given the importance of the NCCP in shaping Canada's mitigation response, we provide more detail on its structure and role. As noted, the NCCP provides overall leadership and guidance in the national response to the climate change issue. It is managed by two key groups: the National Climate Change Secretariat and the National Air Issues Coordinating Committee on Climate Change.[26] The latter committee provides advice to energy and environment ministers and guides the overall direction of the NCCP, including analysis. This committee is composed of civil servants who are mainly from the energy and environment ministries of the provincial, territorial, and federal governments. The National Climate Change Secretariat is the executive arm of the NCCP.

The NCCP has been responsible for assessing the social, economic, and environmental implications of policies and programs in order to develop the National Implementation Strategy. For this task the NCCP created, in the spring of 1998, numerous sector- and issue-based working groups, known as Issue Tables, as part of a National Engagement Process to provide advice, obtain information, and assess implementation options available to Canada to reduce GHG emissions in order to meet a Kyoto-based target. The 16 Issue Tables were comprised of experts from industry, government, academia, and nongovernmental organizations. Over the subsequent two years, the Issue Tables outlined various alternatives and avenues of potential emissions reduction in *Options Papers*.[27] The Analysis and Modelling Group (AMG) consolidates and integrates different analyses in that process.

While there are advantages to splitting tasks among different groups, each addressing a particular facet of the problem, such compartmentalization can hinder interaction and integration. For example, if an Issue Table's proposed option affects energy demand, then energy supply quantities and prices are affected. These effects will in turn change the impacts of other proposed options. Also, changing the process and technological structure of a sector has more than just GHG and energy impacts. Local employment, education, ancillary health benefits, and general social welfare are all affected. Interaction between mitigation options can occur not only at the level of energy supply and demand but also where changes in the costs of products and commodities may affect their market value and their eventual demand.

Much of the analysis in this book issues from this national process of assessing GHG reduction actions and their costs. A key undertaking in this exercise is the definition of costs (Chapter 2) and the application of an integrated model that captures the multiple interactive effects of many actions undertaken simultaneously (Chapter 3).

2
The Challenges of Estimating Emission Reduction Costs

The Challenge of Environmental Policy Making

Canadians care about the environment. In opinion polls they frequently express their desire to protect it. However, most polls about the environment rarely require respondents to assess their desire for environmental protection in terms of how much they are willing to sacrifice of their own income or lifestyle, of how much an incremental improvement in the environment will cost them personally.

This is a problem for public policy. Politicians may earnestly seek to satisfy the public's desire for environmental protection; however, when they present a policy proposal, a familiar cycle can unfold. Those who will bear some of the policy's costs become vocal, claiming unfairness, and soon capture media attention. Politicians appear defensive, inflexible, even mean-spirited. Support for the policy erodes, leaving politicians looking for a way out.

Recent air quality initiatives in the Lower Mainland of British Columbia provide an example. Over the past decade, the Greater Vancouver Regional District (GVRD) developed targets and plans to reduce local air pollution and GHG emissions. Reduction targets were set on the basis of extensive polling and public involvement. The plans included expansion of public transit, paid for in part from annual fees on private vehicles. Municipal politicians, among others, seemed to agree with the policy. Indeed, the regional transportation authority – Translink – even designed a specific automobile fee. However, the announcement of the fee in 2000 raised an immediate outcry from various groups. Although part of this outcry related to the specific context of decision making in the GVRD, some groups protested that they would be unfairly burdened by the fee: municipalities that claim they are not well served by the transit system, the automobile users' association, and vehicle users who argue that they have few options. Before long the media turned its attention to people least able to absorb the extra cost, although even for these individuals the projected fee of up to $120 is about 10% of annual insurance costs and generally under 3% of the total annual costs of owning and operating a vehicle. Letters to the editor complained of cruel politicians, and media columnists and editorialists quickly took up the cause. Soon the political consensus crumbled. Environmental concerns and objectives vanished from public discussion. Less than a year after its introduction, the policy proposal was abandoned.

This example is illustrative of a pattern common in Canada and other countries for policies motivated in part by environmental concerns. Environmental objectives, and the personal costs of achieving them, tend to be unconnected in the eyes of the public and the media. Even politicians who support stronger environmental protection are aware of the fragile and potentially fickle nature of public and media support. Efforts to do something about the environment can quickly backfire.[1]

The Need to Know More about Costs

In this context of environmental policy making, it is understandable that governments have shown some reluctance to act quickly to reduce GHG emissions. As described in Chapter 1, Canadian GHG reduction programs are mostly in the realm of goodwill, information, and voluntarism. To be able to act more definitively, and meet a Kyoto Protocol-sized reduction target, it is critical to have a better understanding of the necessary actions and how the choice of policy to achieve these actions will impact costs and affect the distribution of who pays (and, ultimately, public acceptance). For Canada to shift to a different GHG path, the environmental objective of GHG emission reductions and an understanding and acceptance of the resulting costs to individuals and businesses need to become better connected.

However, understanding the impacts of policy is not a simple task, particularly since the answers cannot be crafted in unambiguous terms (despite the earnest wishes of politicians).[2] What is a cost? Does it include changes to people's lifestyles? What will the GHG future look like without any additional effort? Will renewable energy technologies that are low cost be available in the near future? Can we buy less expensive emission reductions elsewhere rather than achieve them domestically? All of these queries have no simple answer. Nevertheless, only by understanding these issues, and explicitly taking them into account when making cost estimates, can we move ahead. In this chapter, we explain the key cost elements of GHG mitigation and then describe how they serve as a critical basis for the costing methodology and results that we present in later chapters.

Each section of the chapter focuses on a particular challenge of estimating costs. These challenges include the choice of different reference case assumptions, cost definitions, assumptions about the pace and tendency of technological change, as well as what will occur in the US and internationally. We also explore why an action-by-action analysis of efforts to reduce GHG emissions is misleading and look into the importance of including other types of costs, such as indirect costs to the economy and reductions of other harmful emissions associated with fossil fuel use. And we include a section on the relevance of determining not only a national cost but also how this burden will be shared among sectors and regions. By the end of the chapter, readers will have a richer but unfortunately more complicated appreciation of how cost estimates depend on these different dimensions.

Baseline Assumptions and the Impact of Policy

To estimate the cost of moving society onto a different GHG emissions trajectory, we require an estimate of the trajectory that society is currently on and thus where it is likely to be in a policy target year such as 2010. This introduces the challenge that energy forecasters have grappled with for decades: understanding and estimating the diverse factors that cause energy use to change over time. The only difference is that in this case the required estimate is even broader than that of energy forecasting; it includes the evolution of forestry, agriculture, and urban form and infrastructure, among other things.

The Reference Case

There are several terms used for a preliminary forecast. They include *reference case, business as usual, probable forecast,* and *baseline assumption.* We use the term *reference case* in this book. Uncertainty about the reference case increases as we forecast further into the future, although short-term economic cycles can also lead to wide divergences in a short time. Some analysts argue that one should never produce a single reference case because it will give a false sense of reduced uncertainty. An extension of this strategy is to produce an even number of reference cases in order to hinder decision makers from focusing only on the middle forecast and assuming that uncertainty is somehow less than it really is.[3] The National Climate Change Process has thus far worked from a single reference case, *Canada's Emissions Outlook: An Update,* described in greater detail below.[4]

In Chapter 1 (Figure 1.5), we presented a decomposition equation, the Kaya Identity, that links percentage changes in GHG emissions to percentage changes in separate factors: GHG intensity of energy, energy intensity of the economy, economic output per person and population. Given the range of uncertainties associated with each of these factors, their assumed future values in the reference case have a significant impact on the estimated cost of reaching an emission reduction target. Table 2.1 provides the numbers for an aggregate decomposition equation for Canada for the period 1990-2000 and then as forecast by *Canada's Emissions Outlook: An Update* for the period 2000-10. This provides the reference case for our study, but new emissions forecasts can be accommodated within our results, as we explain in Chapter 4.

Table 2.1

Reference case decomposition equation

	Annual % change				
	GHG / E	E / Q	Q / P	P	GHG
1990-2000	-0.4	-0.4	1.0	1.2	1.4
2000-10	0.0	-1.3	1.4	0.9	1.0

Note: P = population; Q = economic output (GDP); E = energy; G = greenhouse gas emissions.
Source: Based on Analysis and Modelling Group, National Climate Change Process, *Canada's Emissions Outlook: An Update* (Ottawa: Natural Resources Canada, 1999), Tables C-8, C-1, and C-24.

As the table shows, the GHG intensity of energy (GHG/E) is expected to remain the same over the ten-year reference case. This is the result of countervailing trends. For example, while natural gas is projected to increase its share of electricity generation relative to coal, which decreases GHG/E, its share will also grow relative to hydro and nuclear power, which increases GHG/E. The energy intensity factor (E/Q) is expected to decline over the next decade at a faster rate than in the period 1990-2000. Again, several effects are involved. Expectations of rising energy prices (even without new GHG policies) encourage energy efficiency efforts and thus reduce E/Q. In the same direction, some of the government's GHG reduction initiatives of the 1990s are expected to carry over into this decade. In contrast, most of the period 1990-2000 experienced relatively low energy prices, which discourage efficiency investments that would otherwise reduce E/Q. The latter two terms, income (Q/P) and population (P), evolve according to expectations of economic growth, immigration policy, and demographics on paths that differ somewhat from the experience of 1990-2000. The combined effect of these factors results in a smaller annual rate of GHG emissions increase compared with the period 1990-2000, but GHG emissions are nonetheless forecast to continue to grow at 1% per year in the absence of new policies to influence the evolution of technology and behaviour.[5]

Policies and Actions

The intent of policies is to move society onto a different trajectory than the reference case, a trajectory that will meet a GHG emissions reduction target. In Table 2.1, the evolution of population growth and output per capita are considered to be much more difficult for government to influence – economically, socially, politically – than the first two terms of the decomposition equation.[6] Energy efficiency is seen as the primary path for reducing the energy intensity of economic output, although, as noted in Chapter 1, structural and substructural changes will also affect – sometimes significantly – energy intensity. There are various ways of reducing the GHG intensity of energy, including switching away from fossil fuels, switching from high-GHG fossil fuels to low-GHG fossil fuels, and capturing and sequestering GHG emissions.

In designing policies and assessing their costs, we must keep in mind the distinction between an *action* and a *policy*.

An *action* is "a change in equipment acquisition, equipment use rate, lifestyle or resource management practice that changes net GHG emissions from what they otherwise would be."[7] Examples of actions are buying a more efficient or alternative-fuel car (acquisition), using your car less often (use rate), owning fewer or no cars by moving to a better-integrated community or working at home (lifestyle), and changing forestry or agricultural techniques (resource management practices). We can estimate the cost of an action individually or as part of a package (portfolio) of actions. The cost is the incremental change in costs (positive or negative) from undertaking the action(s).

The next question is how to induce businesses and consumers to take actions that reduce GHG emissions from what they otherwise would be. A *policy,* or *policy instrument,* is defined here as "an effort by public authorities to bring about an action." Policy instruments are normally divided into different categories. They include the following categories, although some analysts categorize them differently.

- *Actions by government where it has direct control:* Government controls a significant part of the economy with buildings, employees, vehicle fleets, infrastructure, crown corporations, joint ventures, land and resource management authority, and the allocation of its research and development budget.
- *Efforts to influence the actions of firms and households:* Government may try to influence the actions of firms and households (and perhaps other levels of government or other jurisdictions) by various noncompulsory means. They include information campaigns, advertising, product labelling (e.g., eco-certification, emissions and energy use labels), demonstration projects, encouraging voluntary initiatives, and facilitation.
- *Regulation:* Government can set legal requirements on firms, households, and perhaps other levels of government, with criminal and/or prohibitive financial penalties for noncompliance. Examples include appliance, vehicle, and building standards (on energy use or emissions), land and other resource management codes, and standards for technology (e.g., a renewable portfolio standard in electricity requires that a minimum percentage of electricity is produced by renewable technologies).
- *Fiscal instruments:* Government can change the cost of anything through its taxation and subsidy policies. Subsidies include grants and low-interest loans, while taxation policies could include natural resource taxes and pollution taxes.

- *Assigning or extending property rights:* Government can achieve certain objectives by simply clarifying property rights that will in turn enable private markets to determine resource allocation decisions. For example, defining a citizen's legal right to a certain level of air quality could enable that citizen to take civil action against polluters.

Policy instruments are not identical in their effectiveness, and this discrepancy has important implications for estimating costs. For example, an advertising campaign or voluntary program will induce few actions if most firms and households ignore it. The total financial cost to society of emission reductions from this policy is the sum of the costs of the actions plus the government costs of advertising or operating the voluntary program.

A key term used in the national analysis is *measure*. A measure is defined as the "combination of an action and a policy." An example of a measure is a policy instrument leading to more energy-efficient equipment and appliances. The policy is the information campaign or efficiency regulation, and the action is the shift to more energy-efficient equipment and appliances. While we use this terminology, we caution that there is sometimes confusion, even in the National Climate Change Process, in the use of these three terms – action, policy, and measure.

Assumptions about Financial Costs and Consumers' Preferences

As noted, policy analysis throughout the world has been largely focused on the first two terms in the decomposition equation: changing the GHG intensity of energy (GHG/E) and the energy intensity of the economy (E/Q) from the paths that they would otherwise follow. This involves inducing actions that lead primarily to switching to different energy-using and energy-producing technologies. But what does the switch from one technology to another cost?

While it may seem that the answer to this question is straightforward, this has, in fact, been the most controversial question over the past decade of GHG emission reduction policy analysis. Some researchers argue that achieving our Kyoto commitment will cost Canada (and other Kyoto signatories) almost nothing – even be profitable – while other researchers say that achieving the Kyoto target will incur huge costs, with severe implications for the economy.[8] A key reason for this dichotomy lies in opposing approaches to what is a cost.

One approach, referred to as *bottom-up analysis* or *modelling,* is applied more frequently by engineers, physicists, and environmental advocates. This approach focuses on technologies that produce or consume energy in order to estimate how technological innovation and turnover of the existing equipment stock can lead to different quantities of energy use, relative energy shares, and emission levels. Technologies that provide a similar service (heating, lighting, motive force, mobility) are generally assumed to be perfect substitutes but may have different financial costs (the expenditures required to purchase and operate them – capital and operating costs). Thus, the action of switching from a higher to a lower GHG-emitting technology will result in either higher or lower financial costs. If the financial costs are lower, then the cost of reducing GHG emissions is negative; GHG emission reduction in this case is profitable for the firm or household. If the financial costs are higher, then there is indeed a cost of reducing GHG emissions, which, for a particular action, is the change in financial cost divided by the change in emissions – hence the \$/t CO_2e discussed in Chapter 1. Because capital costs tend to occur in the present, while operating costs occur over the operating life of the equipment, these future costs must be converted to their present values – discounted – for comparison with capital costs. In the bottom-up approach, these future costs are usually discounted at what is called the *social discount rate*.[9] We refer to this analysis as a financial cost analysis because it is almost completely focused on comparing financial costs (in present values) of technologies that are characterized as competing alternatives for providing the same energy service.

As it turns out, there are many technologies available today that appear to be cheaper on this financial cost basis yet have achieved negligible market penetration. Bottom-up analysts calculate the effect of a substantial shift toward these technologies and then, by summing up the savings in financial costs and the reductions in emissions, produce an estimate of the net costs or benefits of a particular target for GHG emission reduction. These studies often show that substantial GHG emission reduction can be profitable or cost relatively little.[10]

However, the bottom-up approach is vigorously criticized by many economists, especially for its assumption that differences in financial costs between technologies (at the social discount rate) are sufficient for estimating the full cost of reducing GHG emissions. Economists argue that technologies may differ in a number of ways that are not represented by the financial analysis focus of the bottom-up approach.[11]

- A new technology may be seen as potentially less reliable and therefore riskier. One might expect some inertia in the diffusion rate of new energy-efficient technologies as early models are debugged and consumer confidence builds. Thus, firms and households may require some financial compensation to acquire new, relatively untested technologies because of the greater risk of failure.[12]
- Most GHG-reducing technologies require higher up-front capital investment in order to reduce operating costs, especially energy costs. However, the future benefits of reduced operating costs are uncertain because energy prices can go up or down. The fixed capital associated with energy

efficiency or fuel-switching investments reduces the investor's ability to respond to future unfavourable conditions (irreversibility), so there may be a value in waiting. The longer the period required to pay back the initial investment, the greater the risk and the greater the potential value in waiting to invest or even never investing.[13] Thus, firms and households may focus more on the difference in initial investment cost between two technologies, worrying that they may not realize the benefits of reduced operating costs that would otherwise offset the higher capital cost of the more energy-efficient technology. The effect is equivalent to applying a much higher discount rate to future operating benefits, and this rate reduces the perceived benefits of energy-efficient technologies.[14]

- Two technologies may appear to provide identical services to an engineer but not to the firm or household. Thus, the consumer may be willing to pay more for one technology than another even when the financial costs are the same because of some perceived quantitative or qualitative benefit. For example, engineering analysis may show that an incandescent light and a compact fluorescent light provide the same lumens of lighting service. However, because the compact fluorescent light may take longer to reach full strength, may not fit into certain fixtures, may not look as attractive, may seem to be less bright, or may provide a less attractive hue, the consumer might value the incandescent light more and feel worse off if required to purchase the compact fluorescent light.[15] Economists call this extra value *consumers' surplus,* the value that a consumer gets above the price actually paid for something. One can argue that the full cost of society's switching to technologies that emit fewer GHGs must include any subsidies required to compensate fully for any lost consumers' surplus. Otherwise, there is a loss of value to consumers that has not been tabulated.
- Finally, not all firms and households face the same costs. For any number of reasons, acquisition, installation, and operation costs can vary by location and type of facility. This heterogeneity means that a simple comparison of financial costs may exaggerate the full market potential for penetration of a technology.[16]

For these reasons, economists are associated with an approach, referred to as *top-down analysis* or *modelling,* that derives cost estimates from market observations of how firms and households respond to changes in energy prices or capital costs. For example, if consumers reduce only marginally their energy consumption after a significant increase in energy prices, then economists assume that alternative, energy-efficient technologies do not provide sufficient financial benefits to offset the losses of consumer value that switching to them would cause. Even under unchanging energy prices, one might expect technological innovation to produce more energy-efficient products that offer consumers financial benefits. If little uptake of such products occurs, however, an economist would again suspect that financial costs do not tell the full story of consumers' preferences and thus potential costs.[17]

This approach is called top-down because early cost estimates by economists did not include technological detail but were estimated at aggregate (top) levels of the economy. However, research by economists is increasingly conducted at a technology-specific level. Doing so blurs the traditional distinction between top-down and bottom-up modelling. Technology-specific research by economists now tries to estimate firm and household preferences by direct and indirect methods. Where some market evidence exists, statistical analysis of this evidence may detect indirectly consumers' values; these are called *revealed preferences.* Where market evidence is lacking, say with a completely new technology, researchers might design experiments in which prospective purchasers are asked to choose between alternatives in ways that elicit their valuation of the new technology; these are called *stated preferences.*[18]

Given that the bottom-up approach only includes financial cost differences, at a social discount rate, while the top-down approach also includes estimated losses in consumers' values, including the effects on costs of high revealed discount rates, it follows that bottom-up models have generally estimated low costs of GHG emission reduction, while top-down models have found the opposite. This discrepancy has led to much debate and confusion among the public and politicians.[19]

Bottom-up modellers have justified their low cost estimates by pointing to concerns with the top-down approach. First, they suggest that the top-down approach to estimating costs is tautological because it assumes that any deviation from current practice must incur a cost. In essence, this implies that consumers' current choices are always economically efficient; any deviation leads to higher costs and thus a less efficient outcome.[20] Second, bottom-up modellers suggest that the top-down approach is static in assuming that responsiveness to price change and preferences in technology will be the same in the future as in the past, in spite of technological innovations and evolving values.[21] Because the climate change challenge will be an issue for perhaps many decades, bottom-up modellers suggest that new technologies and changing preferences will change the historically estimated parameters of top-down models. For example, it is suggested that the responsiveness of energy demand to an increase in the price of gasoline may be very different in 2010, when there are hybrid and fuel cell cars as market options, than in the period 1980-95, when the range of technological and fuel choices for automobile purchasers was much more restricted. When this expanded potential for change is included in the analysis, the costs of GHG emission reduction may be significantly reduced.

Although the debate between top-down and bottom-up approaches is now a couple of decades old, there has been surprisingly little methodological convergence. There is now general acknowledgment from both sides that consumer and firm decisions may change with greater information on the financial benefits of energy efficiency investments; both now suggest a role for government provision of information on energy efficiency. But only recently has there been some development of models that attempt to incorporate both top-down and bottom-up characteristics. This new generation of hybrid models has technological detail but also attempts to include estimates of the effects of revealed and stated consumer preferences on choices of technology.[22]

The model that we have developed and now apply to cost estimation is one of these hybrids in that it is based on the principle that *technologies matter and preferences matter.* We explain in Chapter 3 what this means specifically for the method that we apply to the estimation of GHG emission reduction costs.

Assumptions about the Direction and Rate of Technological Change

It is one thing to agree that technological change matters in estimating GHG emission reduction costs. It is quite another to reach agreement on what the major characteristics of technological change might be in the future and how quickly change might occur. Upon closer inspection one can see that the range of technological options is wide indeed.

Decarbonating Options

Humans are just starting to turn their attention to the need to decarbonate their energy systems. This has not been a concern before. Indeed, just fifteen years ago some developed countries were subsidizing their industries to switch to greater coal use as part of a strategy to reduce dependence on imported oil. What some refer to as the *race to decarbonate* has only just begun.[23] Our focus on decarbonating revolves around the following technological potentials: energy efficiency, nuclear power, renewable energy, and fossil fuels.

Energy Efficiency

The degree to which energy systems will need to decarbonate depends, in the first instance, on how efficient they can become. We often think of energy efficiency in terms of the durable goods familiar to consumers: greater fuel efficiency in vehicles, more efficient electric appliances, higher-efficiency heating and cooling systems (including heat pumps), efficient lighting, and better-insulated buildings. While these are indeed key elements of energy efficiency, they represent only part of the potential. Energy efficiency can be pursued throughout the energy system, including supply technologies, delivery networks, industrial use, and even urban form and infrastructure. One such supply technology is *cogeneration,* which simultaneously produces high-temperature heat and electricity. This technology can achieve system energy efficiency above 80%, far exceeding the efficiency of stand-alone thermal electricity plants. Not only does it have an application potential in industry, but it can also be used in commercial and institutional buildings and even in residential applications such as apartments and townhouses. In the latter applications, *district energy systems* can distribute hot water and steam for heating, or cold water for cooling, to buildings in relatively close proximity.

While cogeneration is currently associated with technologies such as steam boilers and gas turbines, fuel cells also coproduce electricity and heat at very high efficiencies, and with no emissions at the point of consumption if the fuel is hydrogen. Fuel cells and other smaller-scale technologies (photovoltaic solar panels, microturbines) can be situated at or near the final consumer, and, if they are also used to produce electricity (*distributed generation*), they reduce the losses from electricity transmission and distribution, thereby improving total system efficiency.

At a higher level of resolution, even the patterns of urban form and infrastructure can have a significant effect on energy efficiency. Cities that more closely integrate residential and commercial activities, on the one hand, and connect pockets of higher-density living with extensive public transit infrastructure, on the other, can use considerably less energy – with lower GHG emissions – than cities with more disconnected and functionally segregated urban sprawl. A concerted effort in this direction is referred to as *community energy management.*[24]

Nuclear Power

Concerns about carbon have given new hope to the nuclear industry. Its advocates point to the substantial improvements in design that enhance the economic feasibility and safety performance of nuclear plants for generating electricity. However, public acceptance is still a large uncertainty. Even without an accident, general public fears about the risks of nuclear power cause uncertainty in the ultimate costs of siting and securing a nuclear plant. Thus, even with the dramatic technical and economic advances that suggest a much safer and more economically attractive nuclear industry, the actual cost of a nuclear path to decarbonation remains uncertain. This cost is not just a function of parameters issuing from the analyses of physicists, engineers, and economists but also involves broader issues of public perception and concern.

Renewable Energy

The renewable energy industry is poised to flourish in a decarbonating world.

Renewables include solar (photovoltaic or thermal), wind, hydropower, wave, tidal, geothermal, and biomass forms of energy. While hydro is a large part of the conventional energy mix in many countries, the other renewables are associated with new and emerging technologies whose costs may drop rapidly with increased commercialization. Modern wind power, for example, was supported by various policies in the 1980s and saw its costs fall by about 75% over 15 years to become competitive with conventional electricity generation technologies in some locations. Photovoltaic electricity and electricity from new biomass gasification technologies (which fuel microturbines) have similar prospects over the next decade. Biomass (from wood waste, crop residues, or dedicated plantations) can also be converted to both gaseous and liquid fuels, the latter providing an alternative to gasoline, in the form of ethanol, for fuelling internal combustion engines. Like nuclear power, large hydropower faces growing public concerns in developed countries such as Canada because of both its environmental impacts (flooding valley bottoms, disrupting water flows and the movements of migratory fish and mammals) and its social impacts (displacing or disturbing Aboriginal and non-Aboriginal communities). However, small hydropower applications – especially those without environmental and social conflicts – still have a considerable potential in many countries.

The future costs of most of these renewable sources are highly uncertain, especially where emerging technologies are involved. Analysts try to project future costs by estimating *learning curves* for new technologies; these are curves that use past experiences to estimate how costs of new technologies may decrease with increasing commercialization and technological development.[25] For example, in the case of nongasoline-fuelled vehicles, costs may decline faster than anticipated. Initiatives to reduce vehicle emissions in the United States (at the federal level, as well as within California) will likely expand the market for clean-fuelled vehicles, potentially reducing costs. Furthermore, government investment may increasingly favour fuel switching over efficiency improvements. Also, some analysts have proposed sharp cost decreases under the argument that the market for passenger vehicles will be completely transformed upon the introduction of one or a few alternative-fuelled vehicle technologies, flipping the market away from gasoline vehicles and toward these more innovative options. If such a transformation were to occur, then costs would fall to much lower levels even at relatively small market shares.

Fossil Fuels

It is generally assumed that a decarbonated energy system will be one with a dramatically reduced role for fossil fuels. But this assumption is now in question. Technologists have begun to look seriously at the possibilities for separating carbon (and perhaps other harmful by-products) from fossil fuels either before or after combustion. CO_2 can be separated from the flue gases of thermal electricity plants. However, postcombustion separation is a more daunting task when it comes to mobile vehicles. This difficulty has led some researchers to focus on precombustion separation. Oxygen-blown gasification of any fossil fuel can produce a synthetic gas comprised mostly of CO and OH. Various separation techniques (e.g., inorganic membranes) are under development to separate the synthetic gas into a hydrogen stream (H_2) and a CO_2 stream. The H_2 stream is then available for storage and transport for ultimate use in direct combustion or in fuel cells. The CO_2 stream would be almost 100% pure and available for transport to storage sites.

Researchers are seriously considering a number of options for permanently storing the CO_2. It could be pumped into old oil and gas reservoirs, producing reservoirs (as part of oil and gas recovery), coal beds (to displace coal-bed methane), saline aquifers 800 metres or more below the Earth's surface, or the bottom of the sea. Enhanced oil and gas recovery already pumps CO_2 into underground storage, but the total capacity for this method is limited. Deep saline aquifer storage is currently seen as the most promising long-term option. These aquifers are plentiful and are found in sedimentary basins within reasonable distances of many fossil fuel deposits. If all of these technology options operate in concert, then it becomes possible to contemplate a fossil fuel-hydrogen/electric future with minimal environmental harm. However, there are uncertainties about these technologies and the public's acceptance of them, so there is considerable cost uncertainty.

The Rate of Technological Change

As the above discussion suggests, there are several promising options in the race to decarbonate. One particular option may ultimately dominate, but this depends on the relative costs, impacts, and risks that emerge from the current flurry of technological innovation. Public acceptability will be critical.

Not only the direction but also the rate of technological change is highly uncertain.[26] Any economy has inertia to the extent that it takes time before certain vintages of equipment, buildings, and other structures are retired and replaced. Under relatively stable market and policy conditions, the rate of technological change depends on both the normal, technical life span of energy-using equipment, buildings, and infrastructure, and on the willingness of decision makers to adopt new technologies. But periods of rapid technological innovation, economic cycles, and a sudden shift in public preferences can all accelerate the rate of equipment turnover and thus technological change. This is sometimes referred to as *economic obsolescence:* equipment that could still operate effectively for many years is replaced simply because the advantages of new equipment are so great. An example would be the purchase of a new, higher-powered computer even when the existing one is still functioning well.

Figure 2.1

Energy decision-making hierarchy

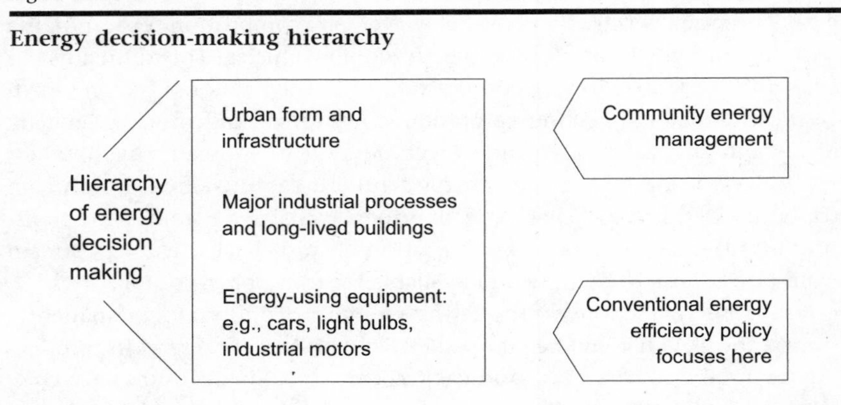

In terms of the expected life span of equipment, the range is wide. Some light bulbs are designed to last less than a year. In contrast, infrastructure can be designed for decades and even centuries. Figure 2.1 depicts, in a decision-making hierarchy, three general categories of energy-using equipment and structures.[27]

The energy-using equipment that we are most familiar with is at the bottom of the hierarchy. Consumer durables such as light bulbs, appliances, and cars can last from one to 20 years as a rule. In the middle of the hierarchy, we place longer-lived equipment and buildings. They include the processing equipment in major industrial plants, electricity generation facilities, and most buildings. They are expected to last from 20 to 50 years and even much longer for many buildings. Urban form and infrastructure are at the top of the hierarchy. It takes decades and even centuries for yearly incremental decisions about land use and infrastructure development to result in profound changes to urban form.

In Figure 2.1 this relationship is depicted as a hierarchy because decisions at higher levels determine or constrain decisions at lower levels. Transportation infrastructure decisions determine the relative use potential for public transit versus automobiles. Urban form (density and location of development and mix of land uses) determines the potential for waste heat recovery, district heating systems, and other cogeneration applications, especially in the commercial/institutional sector. In general, greater energy efficiency is attainable when this objective has been incorporated in decisions at all levels.[28]

At the top level of the hierarchy, the slow rate of change of urban form and infrastructure means that efforts to affect energy efficiency must be undertaken from a long-term perspective – recognizing that incremental changes (say in land use zoning or infrastructure investment) must begin immediately and may not have their full impacts for a long time. This leads to the community energy management perspective discussed earlier as part of energy efficiency. At the lowest level of the hierarchy, it may be possible to delay actions to the extent that opportunities for substantial change through normal equipment turnover arise continually. If there is a need for rapid technological change, then dramatic progress is more likely to be cost effective at the lower levels of the hierarchy. Some people argue, therefore, that today's policies, driven by the long-term risk of climate change, should focus on urban form and infrastructure. These policies can be much more cost effective if they take advantage of natural turnover rates at this level.[29]

Costing Individual Actions versus Integrated Actions

There are many service- and technology-specific studies of GHG emissions reduction potential. There are energy supply experts who focus on cogeneration systems, others on renewables, and still others on nuclear power. On the demand side, there are building experts, appliance experts, and transportation experts, among others. Each industrial sector has its own specialized knowledge about energy use.

As noted in Chapter 1, in Canada's National Climate Change Process, Issue Tables were established to conduct specialized analyses. Each Issue Table produced a substantial list of actions, complete with cost estimates. It is sometimes assumed that a list of actions can be categorized and organized in terms of some common metric, such as cost effectiveness ($/t CO_2e). If they are organized from lowest to highest cost, then they could even be arranged on a GHG emission reduction cost curve. Figure 2.2 depicts a

Figure 2.2

Illustrative GHG emission reduction cost curve

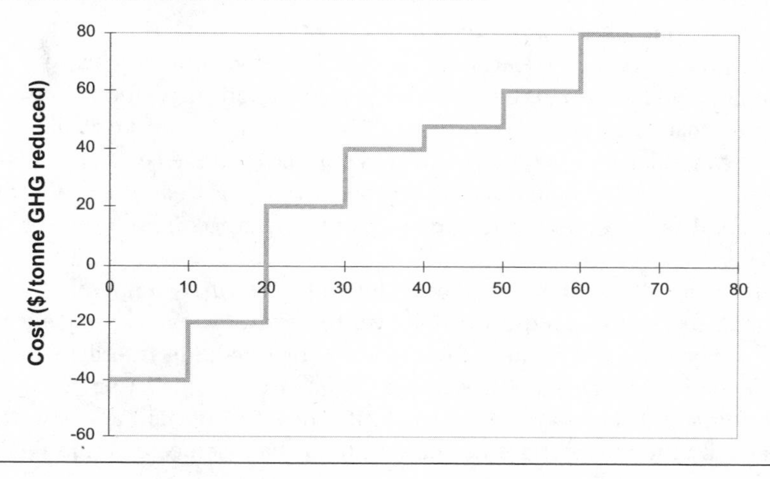

hypothetical cost curve in which the first two actions are negative costs – that is, profitable.

GHG emission reduction costs cannot be estimated, however, by simply adding up the results of individual studies because the costs of individual actions are dependent on which other actions are taken. This is as important at the detailed level within sectors as it is at the general level of interaction between energy supply and demand. Analysis that considers the interaction between actions is required to make coherent GHG emission reduction cost estimates. In a house, for example, the choice of lighting technology affects the amount of waste heat generated by the lighting system. An inefficient lighting system will generate more indoor waste heat, decreasing the required output of the main heating system in winter and increasing the output of the cooling system in summer, if there is one. Differences in the output requirements of a heating or cooling system change the balance of capital and operating costs, which can change the relative cost performances of two competing technologies. Costing analysis needs to take this into account.

When the focus shifts to the level of the entire economy, the issue of integrated analysis becomes critical. For example, there are myriad possible combinations of energy supply technologies for the economy. An analysis of the cost-effective actions for GHG emission reduction for each energy service would require a common set of assumptions about the mix of energy supplies and their prices. However, if one is seeking an outcome that minimizes total costs, then it is necessary to estimate how actions on the demand side will affect the demand for energy supplies, which in turn will affect energy supply decisions and the energy prices facing the demand side. Only by integrating energy supply and demand can one capture the multiple interactions that affect the costs of GHG emission reduction.

Assumptions about Macroeconomic Costs

The issue of feedback relates not just to the interaction of energy supply and demand but also to the possible effects of action costs (and policy costs) on the macroeconomic performance of the economy. In particular, if actions to reduce GHG emissions increase or decrease the costs of providing *intermediate goods* (goods from one producing sector to another) and/or final goods and services, then there may be macroeconomic (or indirect) effects.

In order to estimate these effects, one must first apply a microeconomic, integrated analysis (as described above) to determine the consequences of the actions on relative and absolute costs of providing intermediate and final goods and services. If these cost consequences are small, then the macroeconomic impacts may also be small. If, however, substantial costs are revealed by the microeconomic analysis, then the macroeconomic step is an important component of cost estimation. Likewise, for those who argue that GHG emission reductions will benefit the economy significantly, it is also important to consider macroeconomic impacts. Some claim that profitable emission reductions will provide a *double dividend,* whereby people save financial costs as a result of a policy instrument but also the instrument improves economic performance by applying the revenues from permit auctions or environmental fees to the reduction of distortions in the tax system. However, profitable emission reductions could also lead to a *rebound effect:* an increase in energy consumption that offsets in part the reduction of consumption resulting from an energy efficiency action.[30] This may occur because the efficiency action leads to higher income or reduces the relative cost of the energy service.

A second macroeconomic issue is how domestic producers fare relative to foreign competitors. This is important in the competition for both domestic demand and exports. One can think of this as a vulnerability to cost change. If a Canadian sector has a strong competitive position, then it may be well insulated against the production cost changes resulting from unilateral emission reductions by Canada. If, however, the Canadian sector is in a very competitive market (whether domestic or export), then it is likely to be more vulnerable. A key consideration is whether or not the foreign competitors are subject to the same GHG emission reduction obligations. If this is the case, then the risk of losing market share may be diminished, unless the foreign competitor has some additional advantage in a world where GHG emission reduction is required (e.g., a ready domestic supply of low-cost, non-GHG-emitting energy).

A third macroeconomic issue is the net impact of GHG emission reduction actions and policies on government budgets at all levels. First, the government may have direct financial involvement through some of its GHG policies. This may entail increased revenues from GHG taxes or permit auctions (if applied) and decreased revenues due to tax credits, low-interest loans, subsidies, advertisements, and additional government staff. Second, government will see shifts in its general tax revenues as different sectors of the economy change in relative importance and profitability, and the mix of imports and exports shifts.

The Effect of Policy on Cost Incidence and Total Costs

Cost Incidence

As noted in the macroeconomic discussion above, the costs of GHG emission reduction can vary by sector and region. This is because the direct microeconomic costs and the indirect macroeconomic impacts will not be homogeneous but will depend on the GHG intensities and the particular strengths and vulnerabilities of different regions and sectors. Table 2.2 shows

Table 2.2

GHG intensities (t CO_2e/capita) and the share of emissions produced by sector for each region (1996)

Sector	BC	Alberta	SK	Manitoba	Ontario	Quebec	Atlantic	Canada
Power generation								
GHG per capita	0.50	17.53	13.71	0.00	1.88	0.00	6.17	3.38
% of total GHG in region	3.1%	24.6%	23.7%	0.0%	11.0%	0.0%	31.3%	15.0%
Industrial								
GHG per capita	2.99	8.59	2.94	0.88	5.11	3.84	2.97	4.44
% of total GHG in region	18.5%	12.1%	5.1%	4.3%	29.8%	32.2%	15.0%	19.7%
Residential and agricultural								
GHG per capita	1.50	3.22	2.94	1.76	2.15	1.24	1.93	1.93
% of total GHG in region	9.2%	4.5%	5.1%	8.7%	12.6%	10.3%	9.8%	8.6%
Commercial								
GHG per capita	1.00	1.79	0.98	1.76	0.99	0.69	0.72	1.00
% of total GHG in region	6.2%	2.5%	1.7%	8.7%	5.8%	5.7%	3.6%	4.4%
Fossil fuel industries[a]								
GHG per capita	2.00	22.90	15.67	1.76	0.63	0.00	0.47	3.30
% of total GHG in region	12.3%	32.2%	27.1%	8.7%	3.7%	0.0%	2.4%	14.6%
Transportation								
GHG per capita	6.24	9.30	8.82	6.17	4.66	4.26	5.91	5.51
% of total GHG in region	38.5%	13.1%	15.3%	30.4%	27.2%	35.6%	30.0%	24.5%
Agro-ecosystems								
GHG per capita	0.75	7.51	11.76	7.05	0.99	1.10	0.67	2.17
% of total GHG in region	4.6%	10.6%	20.3%	34.8%	5.8%	9.2%	3.4%	9.6%
Total GHG per capita	16.22	71.19	57.80	20.27	17.13	11.94	19.73	22.54
% of total GHG in region	100%	100%	100%	100%	100%	100%	100%	100%

a Fossil fuel industries include petroleum extraction and natural gas production and processing but not petroleum refining.

Sources: Population data: Statistics Canada, CANSIM [database], Matrix 6547, Catalogue 13-001-XIB. Emissions data: Analysis and Modelling Group, National Climate Change Process, *Canada's Emissions Outlook: An Update* (Ottawa: Natural Resources Canada, 1999), Table C-24.

GHG emission intensities for all regions of Canada by sector, as well as the share of emissions produced by sector for each region. The reduction policies chosen by government(s) can accentuate or offset these regional differences. The cost impact on different sectors or regions is referred to as the *cost incidence* of policies.

The diversity between regions can be understood by focusing on the sharp differences between the economies in Alberta and Quebec with respect to carbon intensity.[31] The Quebec economy has been tied to its hydropower resource for decades, a resource considered to be free of GHG emissions.[32] Almost all electricity in Quebec is generated in this way, and its low cost has resulted in all sectors except transportation being relatively electricity intense. Electricity is used for space heating in the residential and commercial/institutional sectors in Quebec more than in any other province. And, the industrial sector uses electricity for some thermal processes that are normally provided by fossil fuels in other jurisdictions. In contrast, fossil fuels dominate the Alberta economy. Almost all electricity is generated by coal, while most residential and commercial/institutional space heating is provided by natural gas. The industrial sector in Alberta is dominated by fossil-fuel-intensive activities such as oil refining, petrochemical production, natural gas processing, tar sands processing, and coal mining. Thus, in 1996, GHG emissions per capita in Alberta were 71.2 t CO_2e, while in Quebec they were only 11.9 t CO_2e. In percentage terms, transportation has a key share of GHG emissions in Quebec. In Alberta, transportation has a smaller share not because the transportation sector there is any less GHG intense but simply because substantial emissions also come from the oil and gas sector, electricity generation, and space heating.

Different policies will impact these provinces differently. For example, if the federal government initiated an identical carbon tax across the country and used the revenue for a uniform federal expenditure, such as retiring some of the national debt, then the incidence of the policy would be far greater in Alberta. If, however, the government returned to each province the exact amount that it had collected in taxes, then this differential incidence would be mitigated.

The same holds for a policy such as tradable emission permits. If the federal government created GHG emission permits that, by 2010, equalled the annual emission target from the Kyoto agreement, and auctioned these permits to Canadian businesses and consumers, then a much greater per capita amount of the auction revenue would originate in Alberta. The regional impacts would be similar to the original tax proposal. If, however, the government allocated the permits based on a province's share of national emissions at some earlier period (*grandfathering*), then the incidence would be more difficult to determine. If Alberta's incremental cost of reducing emissions is less than those of other regions, then Alberta could actually benefit from the policy. For example, it may be cheaper to lower emissions from electricity production in Alberta when compared with any of the options in other provinces. Some of Alberta's firms would reduce emissions a good deal more than required by their permits, selling the excess permits at a profit to firms in other regions.

Total Costs and the Equi-Marginal Principle

Cost variation by sector and region is important not just for cost incidence but also for determining the total national costs of achieving a reduction target. For example, a requirement that every major sector of the economy reduce its emissions by the same percentage amount below its 1990 levels could lead to different incremental costs (*marginal costs*) for different sectors. At the predetermined allocation, incremental emission reductions may be cheaper in the electricity sector than in the transportation sector, for instance. Total national costs of achieving the Kyoto target could be reduced by having the low-cost sector contribute more than its percentage allocation and the high-cost sector less. In effect, each sector should increase or decrease its emission reduction efforts until it faces the identical incremental cost for the next unit of reduction. (Economists refer to this as the *equi-marginal principle.*) If a participant's incremental emission reduction costs are still less than those of others, then that participant should undertake more emission reduction, while a participant facing higher incremental costs should undertake less. Those taking on more of the burden will see their incremental costs rise until eventually they are equal with those of other sectors. Analysis can determine the savings in total national costs by first estimating the total costs when each sector achieves the same percentage reduction in emissions and then contrasting this estimate with the total costs when sectors are allocated the emission reduction according to the equi-marginal principle.

In brief, the costs of the Kyoto commitment for individual Canadians will depend not just on the actual costs of actions but also on the policies pursued by different levels of government. This critical component of cost estimation is examined in detail in later chapters.

The Full Range of Costs and Benefits of GHG Emission Reduction Policies

The practice of *cost-benefit analysis* involves identifying all of the incremental effects – positive and negative – from a particular policy or set of policies, then discounting and summing them up to arrive at a single net present value. A critical component of this exercise is to estimate monetary values for those effects that are not normally provided in monetary terms, what are called *intangible* or *nonmonetary* costs and benefits. In some cases, these

Figure 2.3

Benefits and costs of GHG emission reductions

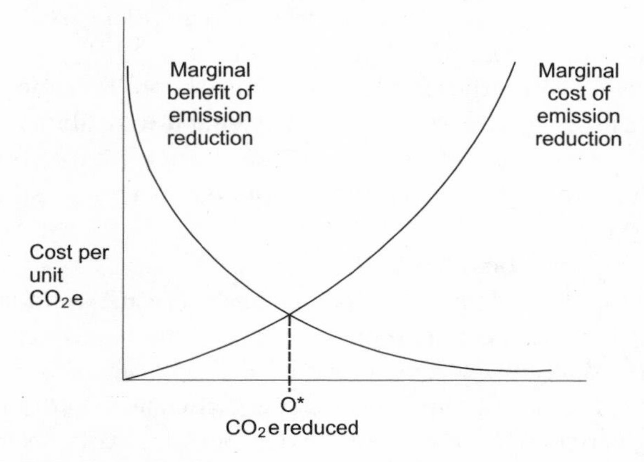

estimates are highly uncertain and controversial in that considerable subjective judgment is required for *monetization.*

Accounting for both costs and benefits of reducing GHG emissions can lead one to determine the optimal level of reduction for Canada or the world. Figure 2.3 illustrates this concept. Emission reductions are on the X-axis and the cost per unit on the Y-axis. The benefits of emission reduction decrease at higher levels of reduction. This is the curve that slopes downward from left to right. The other curve shows the cost of reducing emissions, which increases with higher levels of reduction. The marginal costs of abating a bit further are indicated by points on the line, while the total costs are the full area under the curve at each level of emissions reduction. The point where the two curves intersect suggests the optimal level of emissions reduction (O*). Conceptually, this intersection is the ultimate goal of climate change policy, but uncertainty makes it unclear where the intersection might be.

Cost-Effectiveness Analysis

Where the policy target has already been established (O* in Figure 2.3), analysis can focus on the costs of alternative ways of achieving it. This is called *cost-effectiveness analysis.* The National Climate Change Process in Canada is predicated on the assumption that GHG emission reductions of the magnitude agreed to in the Kyoto Protocol provide net long-term benefits. This is the basis for the analysis presented in this book. Given this assumption, it is not necessary to estimate the benefits of the Kyoto agreement, only the cost effectiveness of alternative actions and policies to achieve it. Thus, in our cost analysis, we have not conducted a financial estimate of the benefits from emissions reduction for either Canada or the world (the downward curve in Figure 2.3). To do so would entail a very different exercise, in both scale and scope.[33]

A related issue is whether net benefits of the Kyoto or some other target would increase or decrease if delays in achieving it were allowed. This issue has generated a great deal of debate; since our focus is on the costs of attaining a near-term reduction target, we do not address this question.[34]

Ancillary Benefits: Local Air Pollution

GHG emission policy can be accompanied by significant secondary impacts. A key example is the decrease in urban air pollution due to cleaner-burning vehicles and reduced use of cars. These air pollutants are frequently referred to as *criteria air contaminants* and include particulates, carbon monoxide (CO), nitrogen oxide (NO_x), sulphur dioxide (SO_2), and volatile organic compounds (VOCs). While these pollutants affect human health, they also affect other animal and plant life (including crops) and damage buildings and structures. If these secondary impacts can be monetized with some confidence, then they could be included in the cost-effectiveness analysis simply as additions to or subtractions from the total costs. If, however, they are complex and uncertain, then one may prefer to characterize them for decision makers in a multiple account framework or even a framework that conveys the range of uncertainty about the ultimate effects. These impacts may be reported in a separate account either in monetary terms, albeit separately, or in physical terms. Analyses that included the benefits of reducing air pollution would show lower costs to the economy than would models that did not include such benefits. We include estimates of these effects alongside our cost estimates.

Assumptions about US Policy and International Agreements

We pointed out earlier that macroeconomic costs depend in part on the position of Canada's industry relative to those of its major competitors. Indeed, decisions made outside Canada affect both microeconomic and macroeconomic costs, which in turn create uncertainty for cost estimation.

It is not known how implementation of the Kyoto Protocol may evolve nor what future international agreements to reduce GHG emissions will look like. We also do not know which countries will ratify the Kyoto Protocol and if it will actually come into force. The US is unlikely to ratify; its position is that any emission reduction agreement must be global and not so ambitious as to cause high costs in the short term.

It is widely believed that international trading will dramatically lower the costs of emission reduction. There are two reasons. First, some developed countries will have lower costs of emission reduction than others, so trading will allow them to find the lowest total cost allocation of responsibility (again, the equi-marginal principle, this time at the international level). Second, if developing countries are included in a trading scheme, then they are assumed to have even lower costs of reduction simply because, as rapidly growing economies, they have much greater flexibility in shaping the development of urban forms, industrial processes, new commercial and residential buildings, and other basic elements of infrastructure.[35]

US policy is especially critical for Canada because of strong trade links. Indeed, just as Canada negotiated a Kyoto reduction target similar to that of the United States, so too Canada will likely choose policy tools that mesh well with those chosen by our neighbour. Therefore, simulating Canada's costs of GHG emission reduction also involves guessing about the goals of the United States and the policy mechanisms that will be favoured by that country for both domestic and international activities. While it refuses to ratify the Kyoto Protocol as currently designed, the US federal government and state governments are enacting various policies to reduce GHG emissions.

As noted in Chapter 1, there is also uncertainty about the full range of actions that will be considered legitimate as part of meeting GHG emission reduction commitments. The definition of a sink and how to treat it have been especially problematic for forested and agricultural lands and the life cycles of forest products. Even with agreement on definitions, verification of compliance will remain a difficult challenge and a continued source of cost uncertainty.

In Chapter 1 we also noted that compliance with the Kyoto agreement by signatory nations would have only a negligible effect on atmospheric concentrations of GHGs. Dramatically greater reductions, probably involving all countries and regions, will be required if stabilization is to occur. This requirement has implications for estimating the costs of achieving the Kyoto Protocol. For example, what should one assume about the emission levels after the target date of 2010? Should one assume that they stay the same? Should they decrease even further? This assumption has substantial implications both for the cost estimates in this study and for the kinds of policies that might be favoured.

Conclusion

This chapter has introduced the many challenges faced when making an estimate of costs to reduce GHG emissions. While cost estimation is not a simple, unambiguous procedure, it is nonetheless a critical step for Canada to move forward with climate change policy. The issues identified in this chapter will resurface in subsequent chapters; indeed, we will refer to them time and again to frame our methodology, results, and policy suggestions. Some issues, such as the importance of incorporating integrative effects, indirect costs, and consumer preferences into a costing methodology, have strongly informed the development of our model, CIMS.[36] We explore these issues further when we describe CIMS in the next chapter. Other issues described here, such as assumptions about the reference case, the future of US and global policy, and the pace and tendency of technological change, relate more to assumptions used in the specific scenarios that we test. We explore these issues in detail in Chapter 4 to understand their relative significance to cost estimates.

3
A Method for Estimating Policy Costs

As discussed in Chapter 2, estimating the costs of policies to reduce GHG emissions is complex and sensitive to assumptions and methodology. Nevertheless, gaining an understanding of how to induce GHG emission reductions in a given country and estimating the consequences on that country's economy and lifestyle are crucial precursors to taking action. This chapter explains the method that we apply for estimating these costs in terms of the list of issues and uncertainties presented in the previous chapter.

Most of our cost estimations are made within the specific context of complying with the Kyoto Protocol, in which Canada agreed to a reduction of GHG emissions to 6% below 1990 levels by 2010. Subsequent international developments, especially the position of the US government, have raised concerns about the future of this agreement. But these developments have no bearing on the method that we present here or on the cost results presented in later chapters, as long as one assumes that Canada intends to reduce its GHG emissions. In our results chapters – especially Chapter 4 – we provide cost estimates for cases in which Canada does less domestically than its initial Kyoto commitment implied. We also develop a national cost curve that shows how Canada's costs change as the target changes, including going beyond the Kyoto obligation. Thus, while the Kyoto commitment provides an organizing framework for our research – especially the detailed results of Chapters 5 and 6 – the future of that particular agreement is not critical to our analysis. As Canada adjusts the amount of emission reduction achieved through domestic actions, the reader can find the appropriate target on the cost curve in Chapter 4 and find on the Y-axis our corresponding estimate of the target's marginal cost (which would be the permit price in a domestic emissions trading program).

In this chapter we first describe various elements and steps of the government-initiated process to determine the costs of achieving GHG emission reductions. We then describe our approach to analyzing costs using the CIMS model. We include sections on the specific structural components of the model, the simulation procedure, model inputs and data requirements, and model outputs. Applying our model in the National Climate Change Process (NCCP) required the use of specific data and methodological steps. We complete this chapter with a description of these data and steps.

The National Process for Estimating Mitigation Costs

Shortly after the Kyoto agreement was reached, Canadian energy and environment ministers at the federal and provincial/territorial levels established a national process to examine the impacts, costs, and benefits of implementing the Kyoto Protocol. This challenging task required substantial work to assimilate large quantities of information in a thorough and structured way and to allow the participation of stakeholders so that the estimates might carry greater legitimacy in shaping policy. With these concerns in mind, the NCCP embarked on an extensive analysis. Options were defined and analyzed by the Issue Tables. These options provided the basis for an integrated analysis to test economy-wide approaches to meeting the target, examine the benefits that Canada might derive from international flexibility mechanisms, identify areas of uncertainty, quantify the potential distribution of regional and sector costs and benefits, and define associated health impacts.

Issue Tables

Sixteen Issue Tables, comprised of approximately 450 government and nongovernment participants, were established to define and analyze the options to reduce GHG emissions. Issue Tables were established both along sector-specific lines – Industry, Transportation, Buildings, Forest Sector, Agriculture and Agri-food, Electricity – and according to cross-cutting issues – Public Education and Outreach, Credit for Early Action, Kyoto Mechanisms, Enhanced Voluntary Actions, Municipalities, Science, Impacts and Adaptation, Sinks, Tradable Permits Working Group, Technology, and the Analysis and Modelling Group (AMG). Each table produced a *Foundation Paper* and an *Options Paper.* Not all of the output fed directly into a cost analysis since only the sector-specific tables and the Municipalities Table (7 of the 16 tables) were instructed to focus on options to achieve at least the 6% reduction (from 1990 levels) within the sector each represented. The AMG was mandated to address analytical needs of the Issue Tables and ultimately to combine their work in an integrated analysis. Early on, the AMG established definitions and parameters for the work of the Issue Tables, including a method for cost calculation. Nevertheless, the content, format, and outputs in the Options Papers were not always uniform and bore the mark of each table's different approach and interests.

Integrated Analysis

The goal of the integrated analysis was to consider the Issue Table outputs with consistency and comprehensiveness in the treatment of time frames, reference case assumptions, definition of costs, and assumptions about the economy's response to cost and price changes. Thus, the isolated analyses of the tables was to be replaced by an integration that included the interplay of energy supply and demand within the economic system as described in Chapter 2. Within this integrated framework, the effects on costs of contrasting targets and policies could be explored:

- sector targets versus a national target
- action-specific policies versus an economy-wide policy instrument.

Total costs and cost incidence (relative impact) depend on the GHG intensities and particular strengths and vulnerabilities of different sectors. As noted in the previous chapter, however, cost incidence can also be accentuated or reduced by the choice of policy target and policy instrument. Five *paths* were established in terms of the reliance on an economy-wide policy (i.e., permit trading, tax) versus an array of action-specific policies (i.e., an appliance efficiency program, renewable energy tax credits, etc.) and on the use of a sector versus national target. Figure 3.1 arranges these

Figure 3.1

Position of policy packages or "paths" in terms of their representation of different attributes

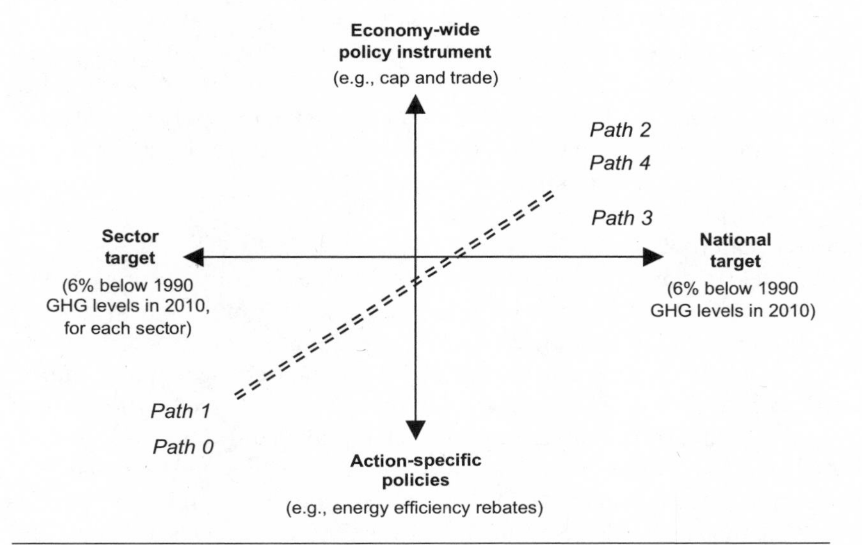

Figure 3.2

Description of policy packages (paths) in the NCCP cost analysis

Path 0:	Aggregation of the measures identified by the Issue Tables in their Options Papers to meet their table targets. Not all sectors achieved the -6% target as directed.
Path 1:	Each sector must achieve an identical reduction from 1990 levels such that the total emissions in 2010 are 6% below 1990 levels.
Path 2:	Sectoral targets are replaced with an economy-wide target of 6% below the 1990 levels, allowing the selection of least-cost measures. Thus, some sectors may exceed -6%, while others fall short.
Path 3:	A cap and tradable permit system is imposed on large final emitters. The sectors not covered by the cap and tradable permit system are affected by measures sufficient to reach a sector target (as in Path 1). Key noncovered emission sources are transportation, residential, commercial, landfills, and oil and gas processing (methane emission for both and CO_2 emissions in oil processing).
Path 4:	The cap and tradable permit system in Path 3 is extended to as broad a system as feasibly possible. Noncovered sources are limited to landfills and oil and gas processing (methane emission for both and CO_2 emissions in oil processing). The noncovered sources are affected by Path 1 measures.

paths according to their coverage of targets and policy types. While policy packages could have also explored the bottom right quadrant of the figure, the NCCP chose to closely relate action-specific policies with a sector target and an economy-wide policy with a national target. Figure 3.2 summarizes these paths. Path 1 requires each sector individually to achieve the Kyoto target of 6% below 1990 levels. Path 2 relaxes the constraint, allowing the entire economy to find the least cost means of meeting the Kyoto target in aggregate (allowing the fuller effect of the equi-marginal principle). Paths 3 and 4 apply a cap and tradable permit policy, although only Path 4 does so in an economy-wide manner.

Analytical Tools

The integrated analysis required a complex and data-intensive process. Several specialized computer models were used, first to simulate how the economic structure, behaviour, and technological makeup of Canadian society would change, and second to simulate how such changes would impact Canada's economy in an aggregate sense. Thus, *integrated microeconomic models* calculated the direct impacts of the paths, and then the results of these models fed *macroeconomic models* and *environment and health models*.

For the integrated microeconomic analysis, the AMG employed two contrasting models: MARKAL and CIMS.[1] MARKAL is a model of the energy-economy system used in many countries and applied in Canada by the Decision Analysis Research Group at McGill University in collaboration with researchers at the University of Montreal. CIMS is the most recent incarnation of the simulation model that we have developed and applied to energy and environment research and policy analysis for the past sixteen years.

Both models contain the same detailed representation of the technologies that produce GHG emissions. Under the influence of a policy package, investment patterns and energy flows change from their reference case evolution. From this change the models calculate the direct costs of the policy packages. For example, if a policy package influences the adoption of greater insulation in houses, then the models calculate the difference in costs relative to the reference case and can thus determine the cost per unit of GHG emission reduction.

Since both models track the turnover of equipment vintages, they might be called bottom-up models. However, MARKAL more closely fits the bottom-up description in Chapter 2 in its emphasis on the role of financial costs in decision making. It is an *optimization model* that informs analysts of which technologies would be selected if all businesses and consumers focused exclusively on minimizing financial costs and had complete information on all other technology choices in present and future. This type of model shows what might happen under these restrictive circumstances. Because of the omission of consumer and firm preferences, however, using such a model to predict the future emission price in a cap and tradable permit policy or the future GHG tax level necessary to meet a reduction target is problematic.[2] In contrast, CIMS incorporates estimates of consumer and producer preferences in its simulation of decision making and in its estimation of the marginal and total costs of GHG emission reduction.

Both microeconomic models focus on GHG-producing technologies and can simulate a substantial share of Canada's anthropogenic GHG emissions. However, some GHG emission reduction options analyzed by the Agriculture Table and the Forest Sector Table were not represented in the CIMS and MARKAL models. These measures were instead incorporated into the analysis by adjusting GHG emissions and costs outside the models.

Because technology decision making by firms and households is part of microeconomics, the AMG labelled MARKAL and CIMS as the microeconomic models in its analytical process. However, both are designed to model changes in the demand for final products and services and in this sense can estimate macroeconomic, structural feedbacks in the economy. But the AMG decided to have the macroeconomic capability of MARKAL and CIMS disabled so that their outputs (changes in investment and operating cost by sector) could be used as inputs by two macroeconomic models: the CaSGEM general equilibrium model of the Canadian Department of Finance and the TIMS macroeconomic model of Informetrica, a Canadian consulting firm.[3] As it turned out, CaSGEM simulated the policy packages of the AMG but did not use the outputs of MARKAL and CIMS. Thus, we report in our results only the aggregate GDP impacts from the TIMS application.

MARKAL and CIMS were therefore applied with industrial output held constant. The exceptions are electricity and coal production, which respond to changes in the demand for electricity from end-use sectors and the demand for coal by the electricity sector. This means that the economy's new equilibrium for each policy occurs after sequential rather than simultaneous simulation of the microeconomic and macroeconomic models.[4]

TIMS assessed how the economic effects of GHG actions ripple through the economy by calculating the direct, indirect, and induced impacts of the policy packages. *Direct impacts* are costs within those sectors directly affected by the policy, while *indirect impacts* trace the changes in spending to their effects on upstream suppliers. Changes in employment, wage rates, and productivity induce further changes in personal consumption and business investment, called *induced impacts*. The outputs of TIMS include changes in economic activity, employment, trade, and government balances.

The NCCP also explored the environmental and health impacts associated with the policy packages and scenario assumptions. A quantitative assessment used micro-modelling outputs that detail changes in energy use to calculate changes in ambient concentrations of air pollutants. The Air Quality Valuation Model (AQVM) of Environment Canada and Health Canada then estimated the physical health and environment impacts associated with the changes in air quality as well as the economic benefits from their reduction. A qualitative assessment accompanied this analysis.

In Chapter 2 we described the methodological debate over the definition and estimation of GHG emission reduction costs. This debate has led to the recent development of hybrid models that combine technological explicitness with firm and household preferences in order to simulate policies to reduce emissions and then estimate the costs. In the case of hybrid models, these cost estimates represent not just the financial costs but also the loss of consumers' value associated with changing technologies from the reference case. CIMS is a hybrid model. In the remainder of this chapter, we explain the way in which CIMS attempts to achieve the goal of bridging the gap between the top-down and bottom-up approaches. This understanding is important for interpreting and situating the cost estimates presented in Chapters 4, 5, and 6.

Design and Simulation Procedure of CIMS

Design

CIMS has been in continuous development since 1986 by the Energy and Materials Research Group in the School of Resource and Environmental

Management at Simon Fraser University.[5] It currently represents seven regions in Canada, but it can be applied to any country or region. As noted above, the model emphasizes the microeconomic level of analysis in that it simulates in considerable detail the equipment selection and building design decisions of firms and households in response to changes in information, costs, and availability of alternatives.[6] However, it can also incorporate indirect feedbacks that are normally associated with macroeconomic models, namely shifts in the demand for final and intermediate products as their costs of production change.

CIMS has been developed and used primarily for energy modelling. Thus, it is structured around the relationships between energy-using sectors (transportation, industry, residential, and commercial/institutional) and energy-producing sectors (electricity generation, fossil fuel supply, oil refining, and natural gas processing). Because it represents technologies in detail, CIMS can simulate any technology-linked material and energy flows in the economy, and it is therefore capable of being a policy tool for a broader range of environmental questions.[7]

Figure 3.3 situates CIMS among the different types of energy-economy models applied to climate policy analysis. The graph has three axes: technology explicitness, preference incorporation, and equilibrium feedbacks.

Figure 3.3

Depiction of energy-economy models in terms of technology explicitness, preference incorporation, and equilibrium feedbacks

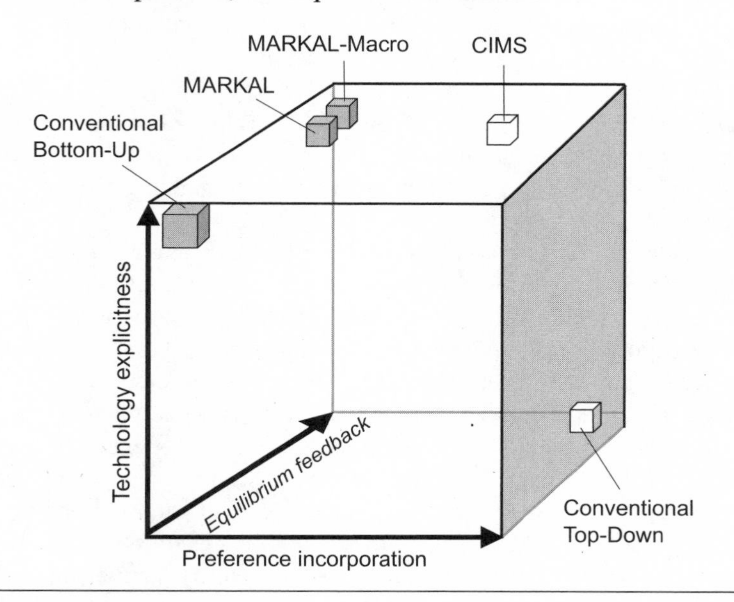

The position of CIMS on the graph distinguishes it from conventional top-down and bottom-up models. A conventional top-down model, especially one with extensive macroeconomic feedbacks, does well on two of the three axes but poorly on technology explicitness.[8] Policies that encourage specific technologies would be difficult to model with this tool. A technology-explicit optimization model, especially one with extensive macroeconomic feedbacks, does well on two of the three axes but poorly on preference incorporation.[9] CIMS is designed to do well on all three axes. It is strongest in its technology explicitness and preference incorporation, and weakest in its macroeconomic feedback. But this feedback is important only when governments are contemplating policies that could have dramatic macroeconomic effects.

The degree to which a model has equilibrium feedbacks is limited in the above presentation to feedbacks within the economic system. Economic impacts would also occur from changes to the climate, just as the climate is impacted by changes in the economy. Some analysts have taken the system analysis to a higher level by connecting economy models with climate models.[10] This type of integrated model could be depicted by changing the scale of the axis related to equilibrium feedback to include climate feedback at the top. In other words, this would be a model in which the economy's performance is a function of climate, which is itself a function of technology choices and behavioural decisions in the growth and evolution of that economy.[11]

Figure 3.4 shows the energy supply, energy demand, and macroeconomic components of CIMS. A central data system (global data structure) keeps track of information flows between the three main components and the interaction between them to reach equilibrium.

- *Energy demand:* The right side of Figure 3.4 includes the industrial, residential, commercial/institutional, and transportation sectors. The industrial component is largest because of its heterogeneous processes and technologies.
- *Energy supply:* The left side of Figure 3.4 includes energy resources and major energy conversion processes (oil refining, natural gas processing, electricity generation). Energy resources (coal, oil, natural gas, renewables) are incorporated as supply curves indicating how cost of production changes at different levels of output, based on estimates of Canadian resources and global markets. Energy conversion processes are modelled using the same simulation process that is applied to energy demand.
- *Macroeconomic feedbacks:* The top of Figure 3.4 includes energy service price elasticities that link changes in the costs of intermediate or final goods and services to their long-run demand. For example, an increase in the cost of production of pulp and paper, caused by purchasing tradable

Figure 3.4

Diagram of CIMS's key components

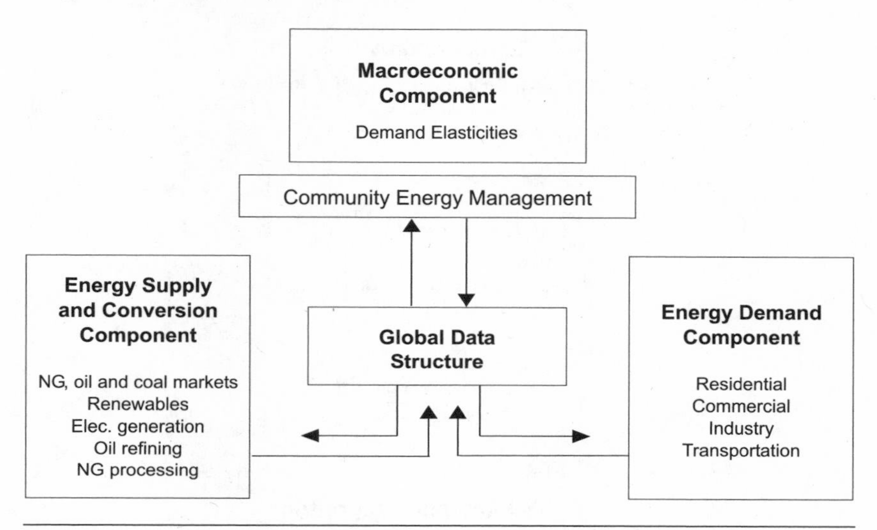

permits for GHG emissions or by investing in GHG abatement, would lead to some reduction in output from that sector if the energy service elasticity has a positive value. A low elasticity suggests that the plant can sustain its sales in the face of cost increases, while a high elasticity suggests the opposite. The macroeconomic feedback is mediated by the community energy management module, which links efforts to influence the evolution of urban form and infrastructure to the level of energy service demand and the opportunities for different types of energy supply. Thus, improving the proximity to work and shopping for some urban apartment dwellers will reduce the demand for mobility in the transportation component, and efforts to cluster commercial buildings will increase the opportunities for cogeneration of electricity and space heat in the electricity component.

Simulation Procedure

The model's simulation procedure can be described in four basic steps, as illustrated in Figure 3.5.

1. *Establish Preliminary Forecast of Energy Service Demands*

   The model represents equipment stocks (including buildings and infrastructure) by the annual quantity of intermediate and final products or services that they provide (person kilometres travelled, square metres of floor space heated, tonnes of newsprint). We call these *service demands* or

Figure 3.5

Basic steps of CIMS

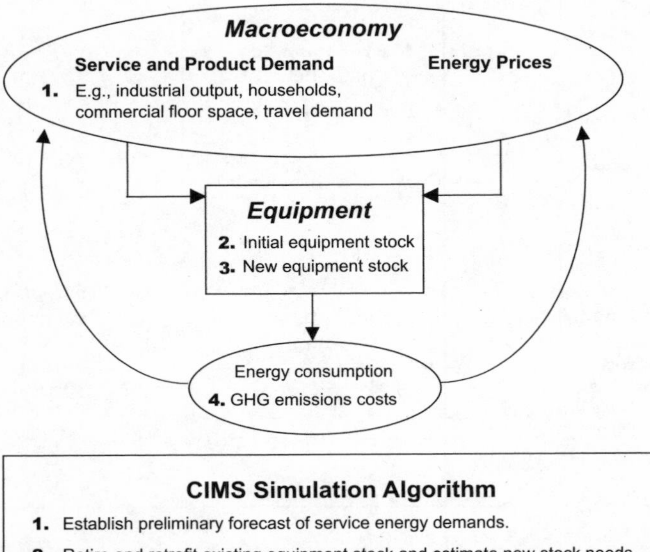

CIMS Simulation Algorithm

1. Establish preliminary forecast of service energy demands.
2. Retire and retrofit existing equipment stock and estimate new stock needs.
3. Determine equipment market shares for new stock.
4. Converge supply, demand, and macroeconomic components.

energy service demands. A reference case forecast of service demands drives the model simulation in five-year increments. If activated, CIMS's macroeconomic component will adjust the reference case service demands as determined by the interaction of cost of production changes with the service demand elasticities (see step 4).[12] Parameters estimated from regression analysis or expert judgment correlate monetary estimates of sectoral economic growth with the energy service demands. This is the critical link between conventional economic indicators ($ of GDP) and the physical indicators (energy services) of interest to bottom-up modellers because it is the evolution of energy services that determines equipment turnover and thus technological evolution.

2. *Retire and Retrofit Existing Equipment and Estimate New Stock Needs*
In each period, a portion of existing equipment is retired. Retirement is time dependent, although a retrofit function may trigger premature retirement or replacement of residual (unretired) equipment because of economic obsolescence.[13] In each time period, the difference between the service demand that can be satisfied by residual equipment stocks

and the forecast service demand determines investment in new equipment stocks.

3. *Determine Equipment Market Shares for New Stocks*
   Technologies compete with each other to capture market shares of the new equipment stocks for each energy service demand. Within a given sector (residential, commercial) or industry subsector (pulp and paper,

Figure 3.6

Energy flow model of the residential sector

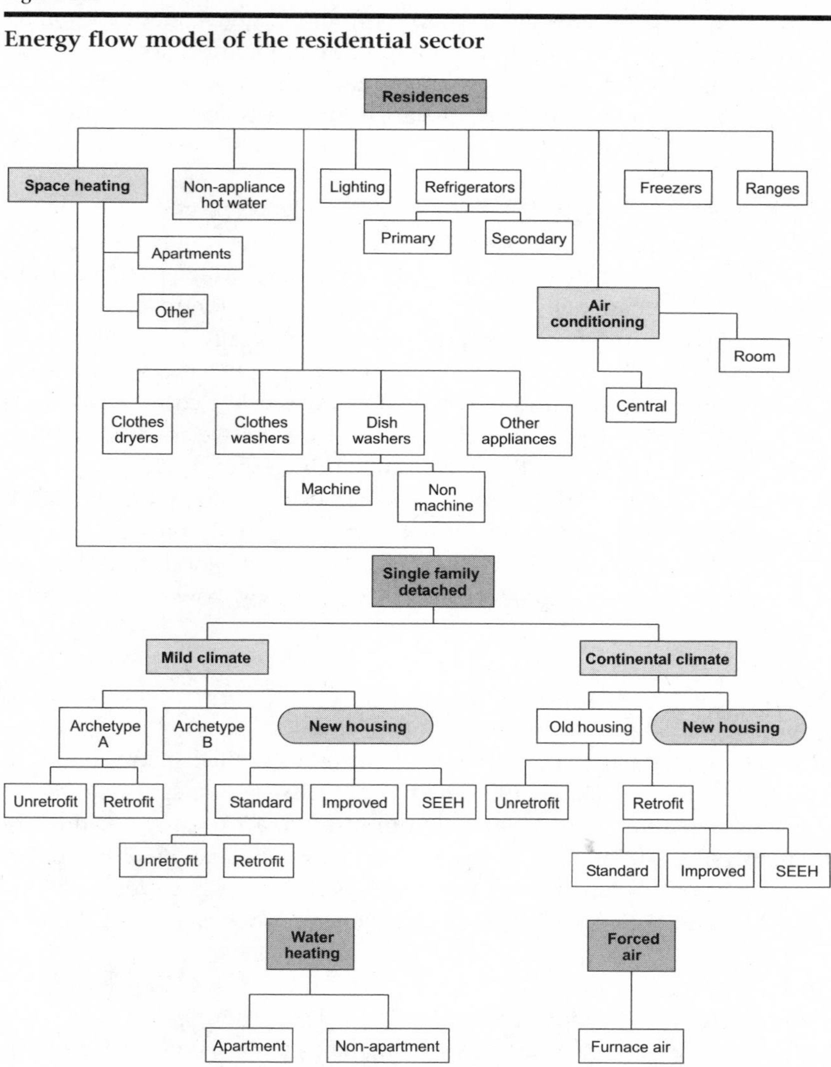

steel), product and energy service demands are linked in a flow model that describes the sequence of activities required to generate that product or service.[14] Thus, the demand for newsprint determines the demand for mechanical pulping, and the demand for residential floor space determines the demand for residential space heating. Figure 3.6 shows the process flow model for the residential sector. The unshaded boxes indicate the energy services where technology competitions take place. Because of the heterogeneity of equipment and processes, major industry subsectors each have their own flow model.[15] In its portrayal of this competition, CIMS differs from a typical bottom-up model and incorporates characteristics of the top-down approach; equipment market shares result from both relative financial costs and the technology-specific preferences of firms and households. A similar method is used for determining the amount of unretired equipment stocks retrofitted in each time period.

4. *Converge Supply, Demand, and Macroeconomic Components*
In each time period, the model iterates between the supply and demand components until energy prices and energy demand have stabilized. In other words, changes in these values from one iteration to the next have fallen below a threshold percentage change – typically 5% – specified by the modeller. The model also cycles through the macroeconomic feedback component (if activated) to see if changes in the costs of providing products and services lead to changes, via the service elasticities, in their demand. Depending on the magnitude of the response, the model may iterate back again through the supply and demand components until it reaches convergence at a new equilibrium. Then the model advances to the next time period. After convergence in all time periods has occurred, the model reports equipment market shares, energy consumption, emissions, and costs.

Model Inputs and Parameters

A model that is both technologically explicit and incorporates preferences at a detailed level requires considerable resources, which may explain in part the limited development of hybrid models. For the specific research reported here, our CIMS team was comprised of 10 researchers, most of them working full time for six months.[16] In this section we describe the key data inputs and behavioural parameters of CIMS. Specific adjustments required for the National Climate Change Process are explained in the next section.

Technical and Financial Characteristics of Equipment

The model contains data on the initial market shares of equipment stocks. Individual types of equipment are characterized in terms of capacity, capital cost, unit energy consumption (and output for energy conversion equipment), non-energy operating cost, emissions, expected life span, and first year of market availability for new technologies. These data are collected from multiple sources, including public statistical agencies, energy utilities, literature reviews, industry associations, and our own surveys of sector experts.[17] The characterizations of existing equipment stocks are derived from diverse surveys and thus have some degree of inaccuracy, especially in terms of the current operating characteristics of older equipment. To deal with this challenge, data on existing stocks are tracked with disaggregated, industry-specific, energy consumption data. Industry and market experts also review the stock estimates.

Inputs and Parameters for Simulating Equipment Acquisition Decisions

In Chapter 2 we outlined the reasons why economists argue that differences in financial costs alone (at the social discount rate) are insufficient for estimating the full cost to firms and households of reducing GHG emissions.[18] We summarize these factors again.

- New technologies are likely to be riskier.
- There may be a value in delaying or avoiding irreversible investments with long payback periods.
- Technologies that are apparent substitutes may differ in terms of nonfinancial consumer values.
- Not all firms and households face the same costs, so market outcomes will be variable.

We also pointed out that economists use various methods of discrete choice analysis to elicit consumer and firm preferences for different technologies. Existing market data can elicit revealed preferences, while direct surveys can elicit stated preferences, the latter being especially important for new technologies or new market conditions. For setting preference parameters in the CIMS model, we use a combination of literature review of discrete choice research, expert judgment, and model validation over a historical period.[19]

The preferences of firms and households with respect to the first two factors (risk of newness and risk of irreversible investments) are both reflected in the effective discount rates revealed from market data.[20] These effective discount rates are indicated by the rates of return that firms and households seem to require before acquiring technologies of higher energy efficiency – most of which are relatively new to the market. This issue has been the subject of intensive research for over two decades, with a noticeable narrowing of the range of estimated discount rates over time. From the

Table 3.1

Discount rates used in the analysis

Sector	Technology	Range	Source	Discount rate
Commercial	Building HVACs	30-50%	Lohani and Azini 1992	20%
	Cogeneration			25%
	Other			30%
Residential	Space heat/shell	26-79%	Hartman and Doane 1986	35%
	Refrigeration	61-108%	Cole and Fuller 1980	65%
	Other appliances	30-70%	Reported in Train 1985	35%
Industrial	Process	20-50%	Hassett and Metcalf 1993	35%
	Auxiliary	>50%	DeCanio 1993	50%
Electricity	Generation		Nyboer 1997	20%
Transportation	Private vehicles		US Department of Energy, NEMS TRAN model	30%
	Buses outside urban areas			12.5%
	Urban public transit			8%

Sources:

Cole, H., and R. Fuller. 1980. *Residential Energy Decision-Making: An Overview with Emphasis on Individual Discount Rates and Responsiveness to Household Income and Prices*. Columbia, MD: Hittman Associates Report.

DeCanio, S.J. 1993. "Barriers within Firms to Energy-Efficient Investments." *Energy Policy* 21, 9: 906-14.

Hartman, R.S., and M.J. Doane. 1986. "Household Discount Rates Revisited." *Energy Journal* 7, 1: 139-48.

literature we have developed a set of typical discount rates that we apply to different types of decisions in different sectors. These rates are presented in Table 3.1.[21] We find that significant changes in discount rates can have a significant effect on CIMS's cost estimates.

The third factor includes consumer and firm preferences that pertain to some intangible character difference between competing technologies. Where we have no indication that a particular technology confers extra value to consumers (over and above the first two factors that relate to time preference), we assume this factor to be zero. However, a review of technology choice research (revealed and stated preferences) reveals cases in which additional attributes give one energy-using technology extra consumer value over its competitors. In most cases the research also provides a monetary estimate of this extra value. We add this monetary value to the financial costs of competing technologies in order to reflect potential loss of consumers' value when switching technologies. For example, gasoline-powered vehicles are still seen by most consumers as having advantages over alternatives when one considers attributes such as refuelling time, range, power, and safety, and research can provide a monetary estimate for this extra value.[22] In a few cases, preference research is scarce, yet experts suggest that if such research were undertaken it would reveal a special value for a particular technology or fuel. For example, a long history of public concern in Montreal about the safety of natural gas space heating in residences appears to have hindered the penetration of that technology in spite of a highly favourable financial position through most of the 1990s. In such cases we simulate the model over a historical period (say 1990-2000) while adjusting a monetary adder that eventually enables the model to track the actual market shares of space-heating technologies over the period.[23]

The fourth factor simply acknowledges that market behaviour is not deterministic. Some firms and households will face (or perceive) different costs than others for the same competing technologies. Discrete choice models reflect this probabilistic nature of decision making with logistic curves that relate the probable market share of a technology to its cost relative to alternatives (and possibly in terms of other decision-maker attributes, such as income level). To simulate this well-known characteristic of real markets, CIMS calculates a single cost per unit of service for each competing technology: the financial costs (capital and operating) are adjusted to reflect any monetized values for additional preferences (factor 3), and the effective discount rates in Table 3.1 (factors 1 and 2) are applied in order to arrive at a net present cost per unit of service. The line at the top of Figure 3.7 shows the results of these calculations. Technology B is more expensive than Technology A. In a simple optimization model, Technology A would capture 100% of new equipment stocks. In CIMS, however, the single-point cost estimates are converted into probability distributions (Wiebull distributions in Figure 3.7), and a probability-based calculation allocates market shares as a function of the degree of overlap of the two distributions. Expert judgment determines the variance applied to the distributions. The variance for industry is usually quite narrow, as on the right side of Figure 3.7, reflecting a greater emphasis on financial cost in technology selection. For some residential services, the variance may be set fairly wide, as on the left side. This means that changes in relative costs will not lead to dramatic changes in relative market shares of competing technologies.[24]

Finally, CIMS has additional levers that allow the user to constrain in various ways the simulation of equipment acquisition decisions. A *maximum market share* can be set for each technology. For example, in the Atlantic region of Canada, the penetration of natural gas technologies is constrained

Figure 3.7

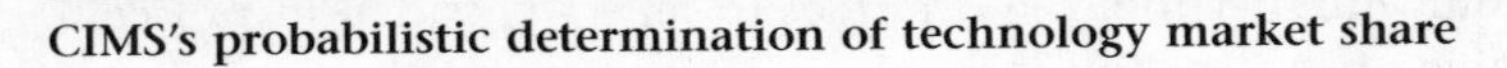

CIMS's probabilistic determination of technology market share

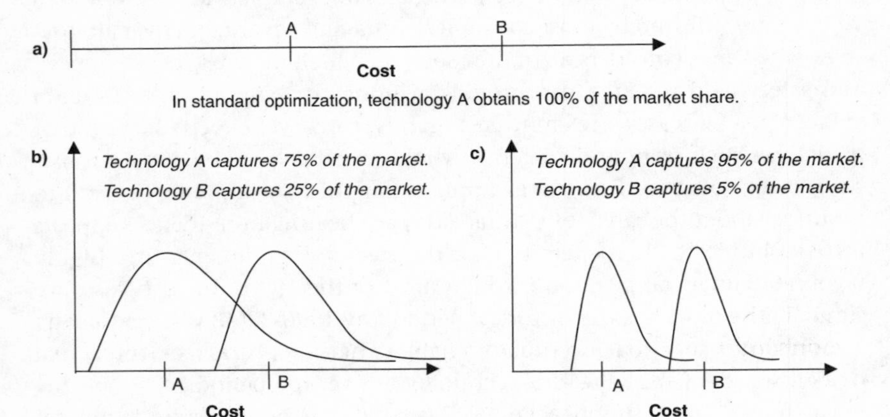

not to exceed the extension of the gas service area in each time period. The service area is rapidly expanding in response to development of offshore oil and gas. Also, the use of natural gas for domestic water heating is constrained not to exceed the use of natural gas for residential space heating. A *minimum market share* can also be set, which is useful for simulating policies that might require minimum equipment sales of a technology that has lower GHG emissions (we describe this type of policy in Chapter 7). The model also has a *declining cost function* that can reduce the cost of a new technology as its market share increases. The user can specify the parameter values of this function in order to reflect how the costs of new technologies decline as their production and use increase.[25]

Driving Variables

For model simulation an initial macroeconomic forecast is required. If it is provided in terms of the monetary value of output (GDP from the cement sector), then it must be converted into the expected growth in energy services (tonnes of cement), as discussed in step 1 of the model simulation procedure. A forecast of international energy prices, notably crude oil, is also required.

Then one or several GHG emission reduction policies are incorporated in terms of their effects on energy prices, investment costs, and technology availability. An example of a single, economy-wide policy is a GHG emission cap and permit-trading system. Another is GHG taxes that lead to higher prices for GHG-emitting energy forms and technologies. Testing alternative tax levels will reveal the unique tax that achieves the GHG emission cap; this should also reflect the price at which permits would be traded. Other policies can be included alongside or to replace the economy-wide policy. For example, equipment subsidies reduce the capital costs of certain technologies, and these can be implemented along with a permit-trading system. A regulation might eliminate a technology from competition or restrict its market share.

Model Outputs: Energy, Emissions, Costs

Energy and Emissions

After equipment stocks have been determined, energy consumption and emissions are calculated and reported based on the unit energy and emission coefficients for each type of technology. CIMS also produces reports on the mix of equipment stocks in each year of the simulation, new stock market shares, and the magnitude and direction of retrofitting.

Costs

When the model is being used for environmental policy simulation, as in the work reported here, the GHG tax or GHG permit-trading price required to motivate sufficient GHG emission reductions represents the marginal cost (financial cost and loss of consumers' value) of the last unit reduced. From this simulation, CIMS then generates two kinds of total cost estimates: strict financial costs, and the total of financial costs and loss of consumers' value. Economists refer to this latter cost as *welfare cost*. As hybrid modellers, we are especially interested in the latter cost. However, the strict financial costs are also of interest to us because they serve as inputs to the TIMS macroeconomic model, which requires information on real financial expenditures on equipment purchase and operating costs. Finally, costs can be disaggregated in terms of the public and private contributions if government subsidies and extra administration costs comprise part of the policy package.

Project-Specific Inputs and Assumptions

The application of CIMS to this GHG costing exercise presented special methodological requirements and challenges. They are discussed below, along with a description of some key inputs.

Reference Case

Figure 3.8 shows the reference case for GHG emissions and Figure 3.9 for energy consumption in the absence of new GHG policies. This forecast is based on *Canada's Emissions Outlook: An Update* (*CEOU*), produced by Natural

Figure 3.8

Reference case, GHG emissions

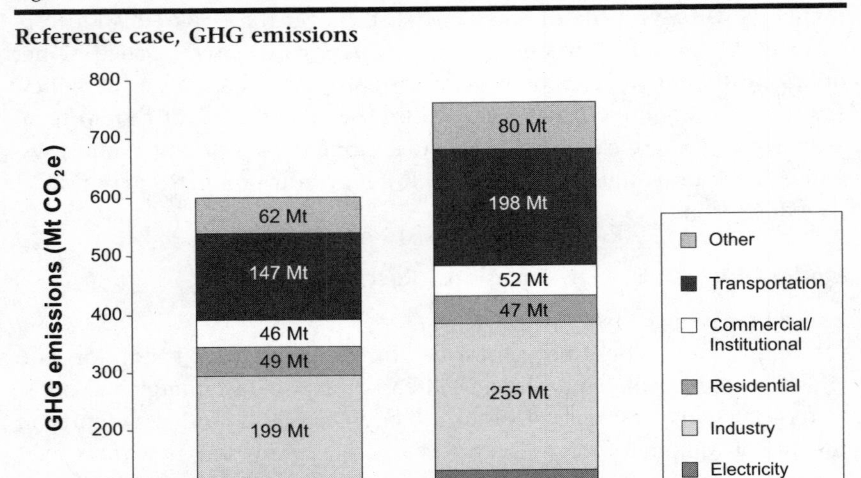

Figure 3.9

Reference case, final energy consumption

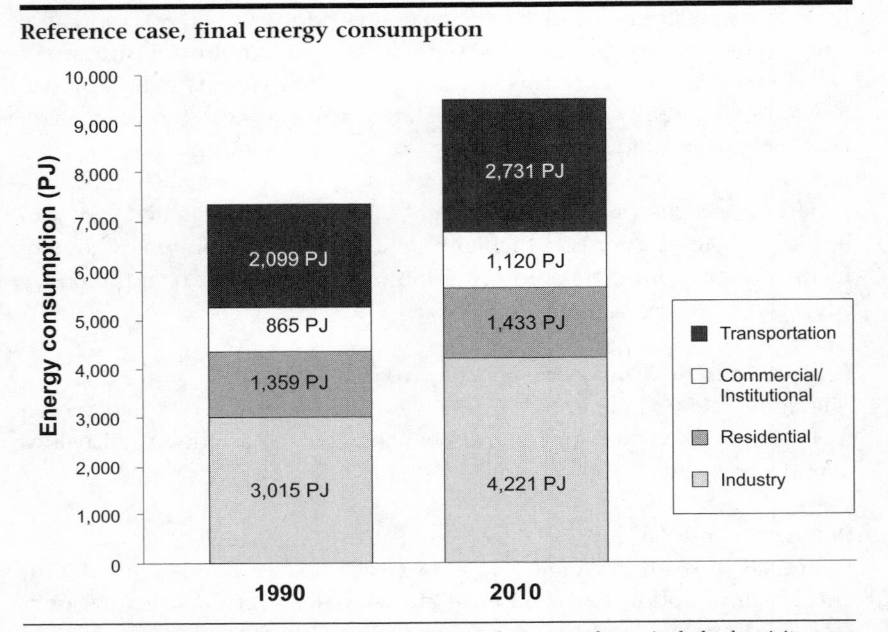

Note: The electricity sector is not represented because the sectors shown include electricity consumption. Electricity sector fuel consumption grows from 2,963 PJ in 1990 to 4,023 PJ in 2010.

Resources Canada. Because CIMS has a different level of disaggregation than that of the models used by Natural Resources Canada, the effort to have CIMS's reference case simulation track *CEOU* involved various adjustments.

Reference case GHG emissions rise from 599 Mt CO_2e in 1990 to 761 Mt CO_2e in 2010, and energy consumption rises from 7.34 TJ to 9.51 TJ. Major growth in emissions and energy occurs especially in the industrial and transportation sectors. For sector-specific details on these GHG trends, see Chapter 5.

Macroeconomic Data and Assumptions

A critical data input in our model is the projection of energy service demands. In the industry sector, projections of physical output by industrial subsector (usually in tonnes) are based on regionally specific growth rates of GDP for 1995 to 2015 provided in *CEOU*. The link between change in GDP and physical growth is not always reliable. While changes in GDP generally indicate changes in physical growth, other industry information guided the development of the actual physical growth rates used in the model. In other sectors service demands were taken directly from *CEOU*, and few adjustments were required.[26] Changes in GHG intensities (t CO_2e/unit service demand) in the reference case are shown in Table 3.2.

Table 3.2

GHG intensities assumed in the reference case

Sector	Unit	1995	2000	2005	2010
Residential	t CO_2e/dwelling	4.25	3.94	3.66	3.36
Transportation	kg CO_2e/pkt	67.30	70.20	70.30	72.10
Commercial/institutional	kg CO_2e/m^2 floorspace	56.10	50.20	43.70	40.00
Electricity	t CO_2e/GJ mWh	0.18	0.19	0.19	0.20
Industry					
Industrial minerals	t CO_2e/tonne output	0.89	0.90	0.91	0.91
Iron and steel	t CO_2e/tonne output	1.39	1.11	1.05	1.02
Metal smelting and refining	t CO_2e/tonne output	3.58	3.03	2.91	2.50
Mineral and metal mining	t CO_2e/tonne output	0.13	0.15	0.15	0.15
Other manufacturing	t CO_2e/tonne output	0.36	0.36	0.36	0.36
Pulp and paper	t CO_2e/tonne output	0.60	0.65	0.64	0.61
Coal mining	t CO_2e/tonne output	33.00	34.70	34.40	34.40
Petroleum extraction and refining	t CO_2e/m^3 output	1.49	1.66	1.76	1.73
Natural gas extraction and refining	t $CO_2e/1000\ m^3$ output	0.36	0.34	0.33	0.32
Chemical products	t CO_2e/tonne output	1.55	0.82	0.81	0.81

In the policy simulations, electricity and thermal coal production adjust to reflect changes in domestic energy demand. However, coal production for export (mostly metallurgical coal) remains unaffected; this is almost all of British Columbia's production and a portion of Alberta's. Likewise, domestic oil and natural gas production remain at reference case levels for all policy simulations, the assumption being that production in excess of falling domestic demand will be exported. The quantities and prices of external energy trade are assumed not to change between the reference case and policy simulations. These are shown in Table 3.3 and Figure 3.10.

Table 3.3

Energy net export assumptions

	2000	2010	2020
Electricity (gWh)	25,736	28,568	17,438
Natural gas (billions of cubic feet)	3,055	3,755	3,755
Petroleum (thousands of barrels per day)	1,055	1,234	1,388

Figure 3.10

Projected energy prices

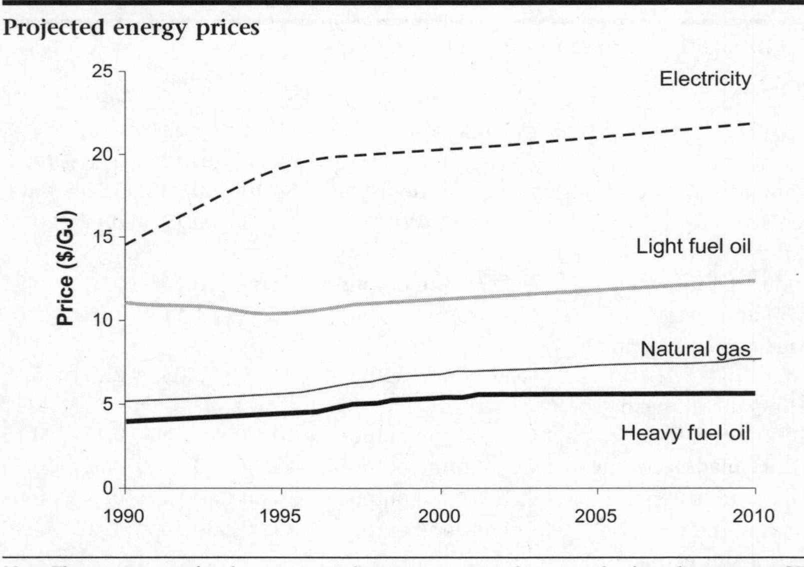

Note: These prices are for the commercial sector; price trends are similar for other sectors. CIMS modelled unique prices for each region.

Table Inputs

The analysis involved incorporating detailed technical and cost information in CIMS from the actions selected by the Issue Tables. Table 3.4 illustrates the types of measures and actions incorporated into the analysis. CIMS's sectors were also redefined to match the definitions used by the Issue Tables (Table 3.5).

Table 3.4

Examples of Issue Table measures

Measure name	Description
National standards for equipment and appliances	This measure involves minimum efficiency standards for a range of products, with new standards introduced in 2004. Labelling would also be included for most products. Products include water heaters, major appliances, furnaces, lighting, and motors.
Government procurement for high-efficiency products	Public agencies and their partners commit to purchasing only equipment that meets qualifying levels of energy efficiency.
Capital infrastructure program for landfill gas capture and flaring	Infrastructure grants are provided for 50% of total project cost to increase landfill gas capture and flaring.
Technology commercialization program (residential sector)	This measure promotes technologies such as advanced lighting technologies, ground source heat pumps, solar water heating, and other proven technologies that have not yet developed a significant market in Canada.
Industrial energy efficiency financial incentives up to \$75/t CO_2e	Financial incentives are provided to industry to invest in technologies that are more energy efficient.
Production credits of 2.5¢/kWh for renewable technologies	Credits are provided to producers based on their electricity output from emerging, renewable technologies. They lower the cost of production relative to the market price.
New vehicle fuel consumption targets	Targets are set for average fuel consumption of manufacturers' fleets.
Enforcement of speed limits	Existing speed limits are enforced to reduce auto and truck speeds.
Transit service	The frequency of bus service is improved, new routes are added, and the safety and convenience of the service are improved.

Table 3.5

Definition of sectors

Sector	Coverage
Electricity	This sector excludes cogeneration activities that occur in the other sectors, such as industry and the commercial institutional sector.
Industry	This sector includes all industry together with mining as well as upstream and downstream oil and gas.
Residential	This sector covers all housing, including attached, detached, and apartments. The Municipality Table's measure affecting land use is partly reflected in this sector.
Commercial/ institutional	This sector includes a wide range of activities and buildings: both institutions such as hospitals and schools and offices and retail stores. Emissions from landfill waste, as reported in *CEOU*, are also included. Most of the Municipality Table's measures are included in this sector.
Transportation	This sector covers all transportation, excluding off-road diesel used in mining, direct process use in industry, and pipelines to transport natural gas and oil. The Municipality Table's measure that changes land use is partly reflected in this sector.

Specific Method for Reporting Costs

Although our study focuses on emissions in 2010, we simulated the economy to 2030. We assumed that policies enacted up to 2010 would continue at the same intensity until 2030. The cost of a path is the difference between its costs and the reference case costs over the period 2000-22.[27] All costs are discounted to 2000, although the currency unit is 1995 Canadian dollars.

Conclusion

The method that we apply for estimating policy costs seeks to clarify and address the major costing issues outlined in Chapter 2.

Our reference case is provided by Natural Resources Canada. While it is prudent to test alternative reference case scenarios, the relatively short time frame between 2000 and 2010 means that reference case uncertainty is much less of a risk to the cost estimates than it would be for a 20- or 30-year horizon.

As a hybrid model that explicitly simulates technological evolution via equipment stock turnover, CIMS combines technology and financial cost information with estimates of the revealed and stated preferences of consumers and firms when facing certain types of decisions. Its cost estimates should therefore be higher than those produced by bottom-up analysis, although – unlike top-down models – CIMS can show how technology-focused policies can reduce costs from what they otherwise might be.

This latter potential can be explored through assumptions about the direction and rate of technological change in terms of both technical innovations and changing consumer acceptance. However, the potential for these dynamic factors to lead to significant cost reductions is restrained by the short time frame between 2000 and 2010.

As an integrated model that includes energy supply and demand, CIMS converges to an equilibrium that ensures that the costs of individual actions are estimated in the context of all other actions that occur simultaneously. This integrated approach can lead to significantly different results than an approach that just estimates the costs of actions on an individual basis because energy prices and other assumptions in this latter case may differ from those issuing from running the integrated model.

While CIMS can estimate some macroeconomic feedbacks, in this study its results provide input to a separate macroeconomic model that generates detailed sectoral and regional impacts.

CIMS tests the effect on total costs and on regional and sectoral cost incidence of alternative target and policy strategies. The alternative target strategies are one single national target versus the same target applied to each sector. The alternative policy strategies are one economy-wide policy instrument versus a large package of policies applied to specific sectors and technology choices.

Additional environmental benefits (especially local air quality improvement) from GHG policies are estimated by separate models and reported alongside our cost estimates. While monetization and integration of these cost estimates with those from the CIMS model may be of interest, it is also important for general understanding to show them separately.

In sensitivity analysis, we test alternative assumptions about US and international policy as well as technology cost and consumer preference changes. These tests are described and their results reported in Chapter 4.

4
National Estimates

Our goal is to look at the challenges of defining costs and their real-life implications. Hence, we consider only briefly the results of the numerous policy packages, or *paths,* that were of specific interest to the NCCP study, and we focus instead on the results in terms of our own costing objectives. We draw in other analyses where useful, both our own and those of others, to help address these objectives. Moreover, we focus on two target/policy approaches that represent the most plausible and illuminating outcomes: Path 1, the sector target that is met with action-specific policies; and Path 4, the national target that is achieved with an economy-wide policy instrument. We refer to them as a *sector target* (Path 1) and a *national target* (Path 4).

Policy Instrument Assumptions

Details of policy design can be especially important to the consideration of money flows and indirect impacts in the economy. In order to explore the real-life implications of the cost results, we therefore define the economy-wide policy instrument in greater detail than set out in the Path 4 definition produced by the NCCP. These details include how the burden (*incidence*) of the economy-wide policy instrument is allocated and assumptions about the policies of Canada's major trading partners.

The policy instrument is assumed to be the *cap and tradable permit* approach. Government sets a cap of total GHG emissions for 2010 that achieves the reduction target and then allocates this cap among participants in the domestic economy in the form of CO_2e emission permits. If the market for permits works effectively, then the permit-trading price gives a common signal to all participants in the economy for comparison with their own estimated costs of emission reduction. The cost of the permits will ultimately be reflected in the cost of energy (or other activities) to the extent that the production and use of energy require the acquisition of emission permits. Thus, the price of gasoline, heating oil, natural gas, and even electricity (if produced by energy sources that create GHG emissions) will increase

Table 4.1

Details of emissions trading coverage

Source and emission	Included?	Source and emission	Included?
Power generation	√	Oil sands	√
Pulp and paper	√	Oil and gas processing (CH_4)	
Iron and steel	√	Air transport	√
Petroleum refining	√	Rail transport	√
Chemicals	√	Road transport	√
Other smelting and refining	√	Residential	√
Cement	√	Commercial	√
Other industries	√	Landfills	
Gas pipelines	√	Other emissions	√
Gas processing (CO_2)	√		
Oil processing (CO_2)			

to reflect the cost of permit acquisition and/or changes in technology. The cap and tradable permit policy covers those sectors of the economy for which its application is administratively practical – almost all sectors (see Table 4.1). The remaining sectors are assumed to be induced to reduce emissions by a mix of other policy instruments, such as technology-specific subsidies and regulations.

When it comes to cost incidence, policy design is critical. The policy instrument can lead to indirect costs that either offset or accentuate the direct costs of reducing emissions. For example, an economy-wide subsidy could be financed by different means of revenue collection, each with different costs to different members and sectors of the economy. Likewise, the revenue from a *GHG tax* could be dealt with by government in many different ways (debt reduction, other tax reductions, increased social program expenditures), each with different costs to different members and sectors of the economy. Regulations tend to impose costs directly on those who are regulated, although if applied to the productive sectors of the economy they could lead to changes in the prices of final goods and services. A tradable permit scheme can *auction* permits or *grandfather* them (allocate permits proportional to initial emissions) or undertake some combination of these approaches. Auctioning is like a GHG tax in that the incidence falls hardest on those with high emissions. In contrast, grandfathering gives more permits to those with high initial emissions; thus, those who have lower costs of emission reduction than the national marginal costs of emission reduction could make money by selling emissions permits even if they have high emissions to start with.

In this study we assume that the cap and tradable permit system is designed so that macroeconomic consequences are minimized, which is achieved by ensuring that the financial signal is provided to the economy in ways that minimize transfers of wealth. The tradable permits are allocated on the basis of emission levels at some initial period, say 2000, although some auctioning of permits helps to offset the dramatic gains that might otherwise be realized by those with the lowest costs of emission reduction. Thus, allocation would be primarily determined by grandfathering, with a small percentage by auction. Changes in costs – and thus total impacts – would therefore be restricted to the additional costs arising from investments in energy efficiency and fuel switching that are motivated by the cost of permits. Only the direct costs of actions (investing in more expensive equipment, shifting to more expensive forestry practices, building extra transit infrastructure) cause a financial burden and macroeconomic effects are therefore minimized.

Finally, we assume that the United States and Canada implement similar policies to reduce emissions. Canada is especially motivated to match its policy with that of the United States, given the importance to our economy of trade with that country. Even though the US is unlikely to implement the Kyoto Protocol as specified in 1997, its federal and state governments are already implementing policies to reduce GHG emissions, and this effort is likely to intensify. With similar policy approaches, the relative cost impacts of emission reductions should not lead to extreme differences from one economy to the other.

Later in this chapter, we present results from a similar cap and tradable permit study of the US economy that uses the main energy policy model of the US government. The study gives cost estimates similar to those in our study, which supports the argument that the indirect impacts of Canadian GHG emission reductions – as committed to at Kyoto – may not be great, at least in terms of Canadian comparative trading advantages with the United States. If, however, the United States halts its pursuit of GHG emission reductions, then the impacts on Canada could be substantial if our domestic policies are not adjusted in ways to minimize direct cost impacts to vulnerable industries.

Simulation Issues and Concepts

As noted in Chapter 3, the results reflect the change in activity relative to a reference case with no additional domestic efforts to reduce emissions. The results are therefore incremental to the activity that would have occurred. Results are typically reported for 2010, the midpoint in the Kyoto commitment period (2008-12). The costs that we report in this and subsequent chapters approximate welfare costs – the total of financial costs and loss of consumers' value.

Under a well-functioning cap and tradable permit system, the permit-trading price corresponds to the country's *marginal cost* of emission reduction. This concept is integral to CIMS's simulation of policy responses. CIMS identifies the unique GHG permit-trading price (or GHG tax) for which the sum of actions in a sector, or across sectors, achieves a reduction target at the lowest possible cost.

By simulating the model repeatedly in achieving increasingly higher targets, we produce a cost curve for Canada: a relationship between the unit cost of GHG emission reduction and the levels of reduction. This cost curve is useful for informing negotiations to revise the Kyoto agreement, create a new agreement, or address a new reference case in which the required emission reduction is different.

In this chapter we present and discuss the national results from CIMS (subsequent chapters provide a more detailed look at regional and sector results). We first present the costs of the full set of policy packages and discuss the insights that they provide. We then focus on two particular policy packages and discuss what they might mean to Canadians in their daily lives, including impacts on the everyday prices that Canadians see and the equipment that they use. Later sections focus on how the cost results compare with other costing analyses and on cost sensitivities to key assumptions and uncertainties.

Policy Package Cost Effectiveness

The analysis was designed so that it would provide insight into the way in which the target could be apportioned – a national target versus a sector target – and how it could be attained – an economy-wide policy versus an action-specific policy. As described in Chapter 2, these issues were expressed as five policy packages by the NCCP, referred to as *paths*.

- Path 0 is an aggregation of the measures identified by the Issue Tables. Each Issue Table was instructed to put forward enough measures to meet an emission target of 6% below 1990 levels.
- Path 1 allows Issue Table measures to go further in inducing actions and for new actions to occur.[1] Sectors still attempt to meet their sector targets.
- In Path 2 sector targets are replaced by a single, national target allowing the selection of the lowest-cost actions across all sectors.
- Path 3 applies a cap and tradable permit system that includes large final emitters (many industrial subsectors and the electricity sector) but excludes the residential, commercial/institutional, and transportation sectors from the trading system.
- Path 4 represents a cap and tradable permit system whose application is as close to economy-wide as feasible.

The results show that the paths with sector targets are unable to bring overall national emissions to 6% below 1990 levels in the period between

2008 and 2012. As illustrated in Figure 4.1, both of these policy packages, Path 0 and Path 1, fail to make enough reductions to bring national emissions to 564.5 Mt CO_2e in 2010 (Canada's Kyoto target). Path 0, the aggregate of the Issue Table measures within an integrated model, misses the target by 7%, while Path 1 misses it by 5%. Although the Issue Tables were instructed to put forward enough actions to reach 6% below 1990 levels for every sector, this goal was not achieved by all. Despite the fact that in Path 1 more flexibility is available to meet the sector target than simply the aggregate of the Issue Table measures, the industrial sector still does not include enough economically feasible actions to meet its target. This lack is due to the substantial growth expected to occur in industrial emissions in the reference case between 1990 and 2010 (28%) and the fact that total industrial production is not allowed to decrease (initial macroeconomic adjustments are prevented). Meeting a sector target requires a total reduction of approximately 64 Mt CO_2e from industry, more than all GHG emissions currently produced by residences in Canada. In reducing emissions beyond 37 Mt CO_2e, the level of emission reduction attained in Path 1, lower-cost actions are used up, and additional options either do not exist or are prohibitively expensive.

Figure 4.2 shows total costs for each package, while Table 4.2 puts the costs in context with emissions reduced. When each sector must meet a proportional share of the national target, Path 1, the overall costs are high.[2] Costs decline when investment in energy-producing and energy-using equipment is allowed to reconfigure itself in a least-cost manner across sectors.

Figure 4.1

Emission reductions by path in 2010

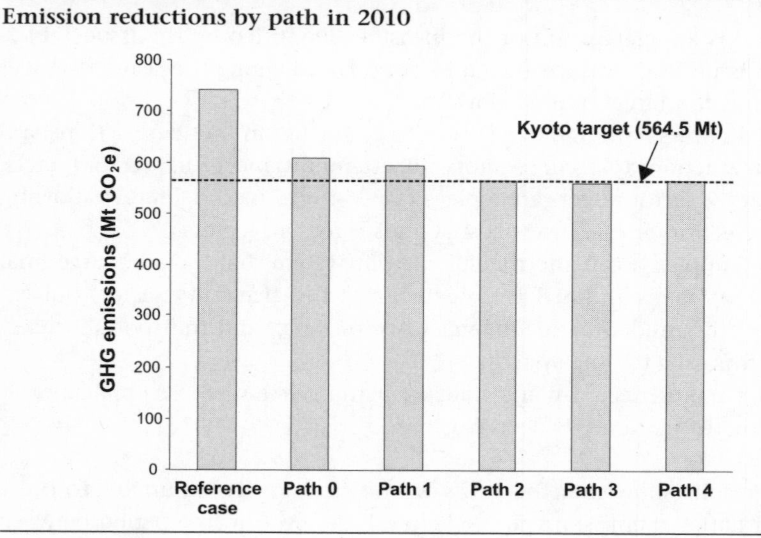

Figure 4.2

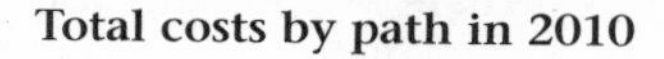

Total costs by path in 2010

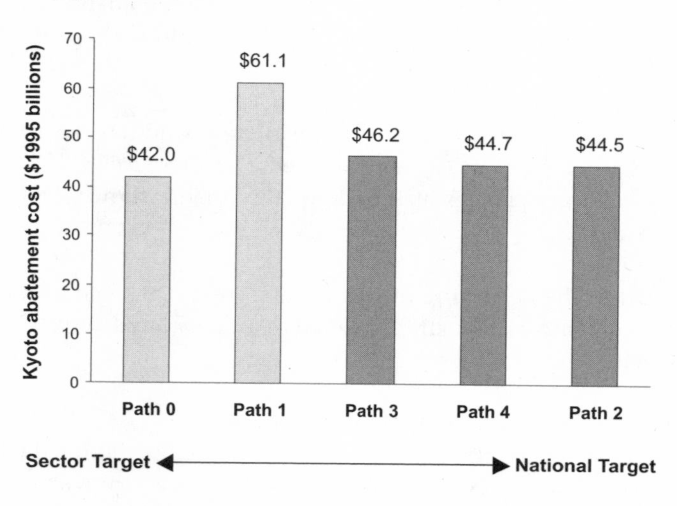

Table 4.2

Total costs and emissions reduced, by path

	Total cost (billions $1995)	Emissions reduced (Mt CO_2e in 2010)
Path 0	42	135
Path 1	61	150
Path 3	46	180
Path 4	45	182
Path 2	44	181

This reconfiguration occurs with domestic permit trading, in which those firms with lower costs of emission reduction do more than their share and sell some of their permits to those who have higher costs. The outcome supports the equi-marginal principle that moving from individual sector targets to a national target achieves the desired objective at lower costs.

What These Outcomes Would Mean for Canadians

The results show that meeting the Kyoto target would cause direct costs of $45 to $60 billion. But what do these costs mean? Is a total direct cost of $45 billion enough to devastate the economy, or is this an amount that can be taken in stride as the economy grows? What might happen to the competitive position of Canada, an export-oriented nation? More fundamentally, what do these results suggest about how the daily lives of Canadians may be affected? Will fuel costs become exorbitant? Will there be dramatic job losses? What about other considerations such as air quality and health? These issues are the focus of this section. We address them through the following key questions.

- What are the changes in society, infrastructure, and technology that will enable Canada to meet the emission reduction target?
- How will these changes affect the quality of life/standard of living of Canadians?

As noted at the beginning of this chapter, from this section forward we focus only on Path 1 and Path 4, referred to as *sector target* and *national target* respectively.

Changes that Bring about Emission Reductions

Many popular culture images of the future foresee a world filled with fancy bubble cars, solar panel roof tiles, and personal battery packs. Read any science fiction work and it may well describe some technological change that will impact energy use and GHGs. Nevertheless, the changes in society that will enable Canada to meet the Kyoto target in the near future are not particularly radical and in most cases are simply a stronger penetration of equipment already in the marketplace. Key sources of emission reductions are described below. Richer detail for each sector can be found in the next chapter.

Before focusing on the specific changes that need to occur to reduce emissions, it is useful to step back and consider the aggregate changes to the terms of the Kaya Identity for the national target, Path 4 (Table 4.3). This decomposition equation, introduced in Figure 1.5, breaks down percentage changes in GHG emissions as the sum of percentage changes in the GHG intensity of energy, energy intensity of the economy, total economic output per capita, and population. At a broad level, emission reductions are accomplished by reducing at least one of these ratios, and the identity gives a rough idea of the magnitude of change required to meet the Kyoto target. Since population is not subject to mitigation efforts, only the first three terms change in the equation (the first three columns in Table 4.3). For the policy simulation, the change in energy intensity (E/Q) is stronger than the change in the GHG intensity of energy (GHG/E). Change in Q/P indicates a different rate of economic growth.

As depicted in Figure 4.3, improvement in energy intensity is the most important type of action, but fuel switching, particularly in the electricity sector, is also important. In addition to energy efficiency in equipment, energy intensity in this figure represents reduced energy for transportation,

Table 4.3

Decomposition equation applied to the results

	Simulation	% change				
		GHG/E	E/Q	Q/P	P	GHG
1990-2000	Reference case	-0.4	-0.4	1.0	1.2	1.4
	National target	-0.4	-0.4	1.0	1.2	1.4
2000-10	Reference case	0.0	-1.3	1.4	0.9	1.0
	National target	-1.6	-2.7	1.4	0.9	-2.0

Notes: P = population, Q = economic output (GDP), E = energy, GHG = greenhouse gas emissions.
The reference case is from CIMS (which is similar but not identical to the *CEOU* reference case described in the decomposition equation in Table 2.1). The national target rate of GHG reduction (-2.0% per year) is calculated from the CIMS simulation and thus does not include the reduction in GDP growth estimated by the Informetrica model. If this were included, the 1.4% growth in Q/P would decrease to 1.1%, and the -2.0% change in GHG becomes -2.3%.

including using less fuel per passenger kilometre travelled and less demand for travel. The *sinks* category in Figure 4.3 indicates the share of the emission reduction target that the analysis assumes will be met by including sinks from current forestry and agricultural activities. The *other* category refers to actions directed at preventing *fugitive emissions,* such as those from landfills, oil and gas pipeline leaks, and livestock production, capturing and sequestering CO_2, and augmenting sinks through land-use practices. The *other* category is reflected in the GHG intensity term in the Kaya Identity shown in Table 4.3.

Figure 4.3

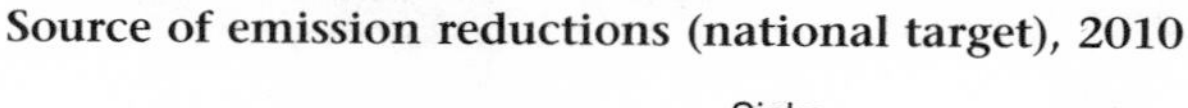

Source of emission reductions (national target), 2010

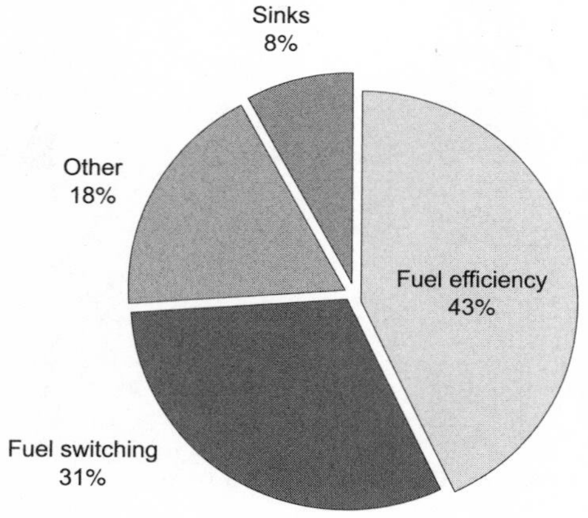

Clearly, changes in the ways that our society uses energy lead to the majority of emission reductions. This is not surprising since most anthropogenic GHG emissions are energy based. By 2010 not only will we use less energy, but we will also move to less carbon-based fuels. Figure 4.4 shows the changes in fuel types that lead to direct emission reductions (emissions reduced directly by end-use sectors). Both policy simulations show a similar pattern: energy consumption drops by approximately 15% through efficiency improvements and reductions in energy demand, while fuel switching occurs from refined petroleum products to natural gas and electricity. Even though less total energy and fossil fuel are consumed in the sector target simulation, this reduction does not lead to fewer overall emissions, and this simulation does not meet the emissions target. This is because end-use energy tells only part of the story; it is also necessary to consider changes to the generation of electricity. Figure 4.5 indicates that *indirect* emission reductions in the electricity sector occur primarily via fuel switching from more carbon-intensive fuels such as coal and oil to hydroelectricity and other renewables, although there are also improvements in the efficiency of generation. The efficiency is shown by relating the *electricity generated* bar in the figure to the total of fuels consumed. Efficiency increases between the reference case and the national target from 58% to 63%. The electricity sector results are discussed in more detail in Chapter 5.

Figure 4.4

Energy consumption, 2010

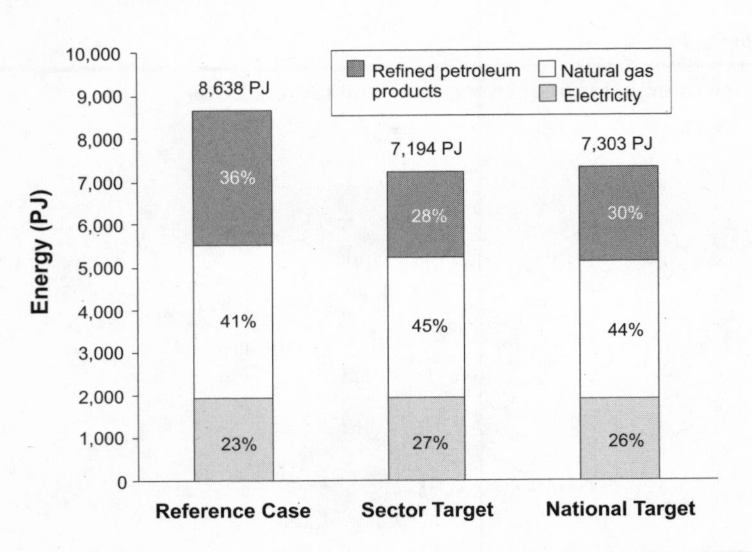

Figure 4.5

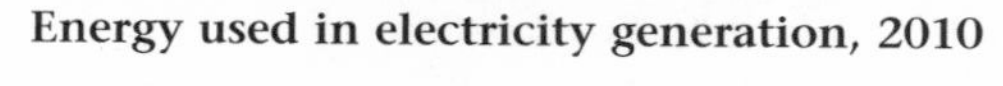

Energy used in electricity generation, 2010

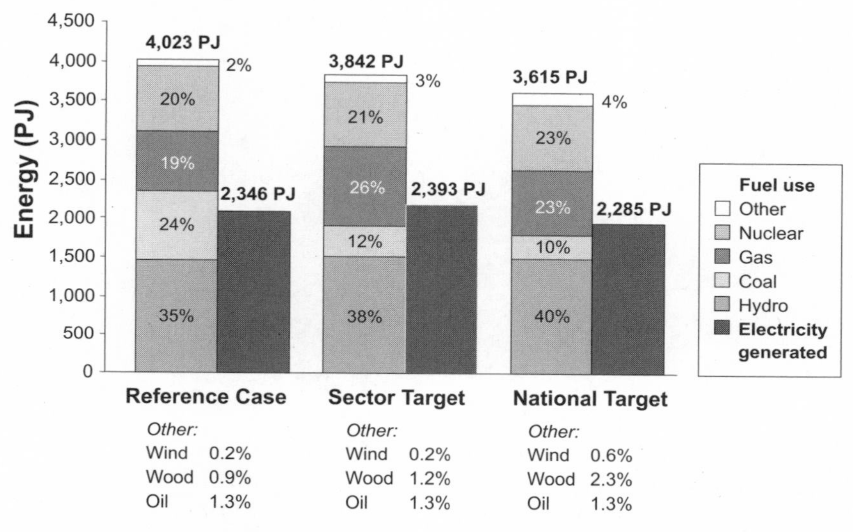

Figure 4.6 shows that under the national target most emission reductions occur from changes in electricity generation, transportation, and industry. Under a sector target, transportation contributes relatively more and industry and electricity less. The significance, cost, and source of emission reductions for each sector under both targets are described in detail in the next chapter. The reductions themselves result from a combination of hundreds of technological changes across all sectors. A few key actions and sources are highlighted below. All figures refer to emission reductions under the national target.

- A reduction of 27 Mt CO_2e occurs through the capture and sequestration of CO_2 in Alberta and Saskatchewan. CO_2 is stripped from combustion flue gases in the generation of electricity and piped into oil and gas reservoirs and into deep saline aquifers. This technology is new and relatively untested.
- A reduction of 43 Mt CO_2e occurs when utilities switch from coal to natural gas and renewables for electricity generation, particularly in Ontario and the Atlantic provinces, where capture and sequestration of CO_2 is not possible in the 2010 time frame. Nationally, coal use declines significantly from 24% of fuel used in generation to 10%. This decline happens despite the fact that capture and sequestration of CO_2 can allow for continued

Figure 4.6

Emission reduction by sector (national target), 2010

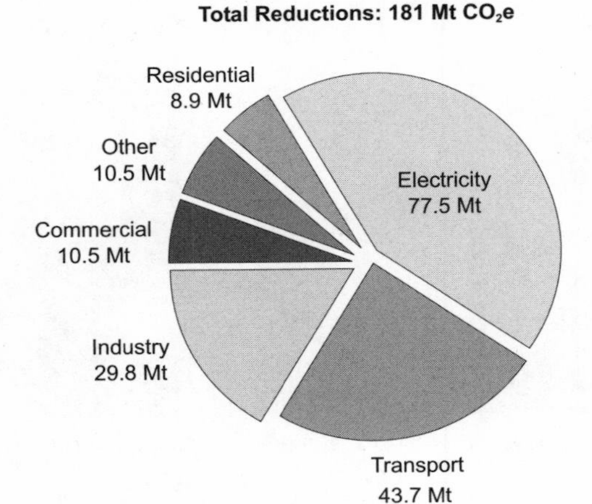

coal use. Switching to natural gas is in many cases cheaper than capturing and storing CO_2, especially at lower permit prices.

- A reduction of 10 Mt CO_2e occurs from shifting some private vehicle use to transit by creating greater pricing incentives to use transit. Transit service is enhanced by increased infrastructure, improved operation, and better coordination of land use planning with transit infrastructure (concentrated development near transit corridors).
- Actions in the oil and natural gas industries reduce 18 Mt CO_2e. This occurs by reducing emissions from combustion and by reducing the amount of fugitive emissions.
- A 6 Mt CO_2e reduction occurs from collecting landfill gas from municipal sites through a series of wells and piping systems. The gas is either flared or used to produce electricity, which can generate further reductions depending on the fuel that it replaces.

Quality of Life/Standard of Living

Changes described in the preceding section influence the way that our communities and economy develop, ultimately affecting the standard of living of individuals. Although GDP is not a comprehensive indicator of material well-being, changes in GDP are a common way of measuring the impacts of acting on climate change. An increase in per capita GDP suggests an improvement to quality of life, but other factors must be considered: for instance, economic growth can be associated with higher infrastructure costs, a decline in air quality, and traffic congestion. To assess how meeting the emission reduction target will affect standard of living, we look beyond economic growth impacts to include other values important to individuals: family income, price levels, community infrastructure and services, air quality, and other environmental attributes.

According to the CIMS-based macroanalysis undertaken for the NCCP, meeting a GHG reduction target such as that of the Kyoto Protocol will impact Canadian economic growth and the productivity and input costs of some businesses negatively, although this impact will vary across sectors and regions. While this economic impact is important and should inform policy design, a look at other elements that contribute to quality of life suggests that meeting the target will

- have a relatively minor negative impact on family incomes
- be accompanied by co-benefits that improve quality of life and allow for more sustainable communities
- have limited lifestyle impacts.

Impact on Economic Output and Businesses

The implementation of actions to meet the reduction target is expected to have a negative impact on economic output. Informetrica's macroeconomic analysis determined, based on CIMS's direct total cost and energy consumption results for the national target, that cumulative GDP will be reduced by about 3% by 2010. The maximum net loss in annual economic output is estimated to be approximately $27 billion in 2010.[3] To meet the target deadline of 2010, increased investment in emission-reducing equipment initially stimulates economic activity. Thereafter, higher production costs, deterioration in productivity, and lower incomes combine to reduce GDP to below reference case levels. As noted in Chapter 3, this estimate does not include the benefits of reducing the threat of climate change, nor does it include nonmarket co-benefits such as reduced local air pollution.[4]

A cumulative loss of 3% of GDP by 2010 means that over the decade, if the economy would normally have grown by 30% (as is expected), it will instead grow by about 27%.[5] When one considers the large risks involved in climate change, the loss of a year's growth in economic output – the equivalent of a year's recession – may not seem to be onerous. Nevertheless, perspectives differ. John Bennett, a climate change specialist at the Sierra Club of Canada, told the *Globe and Mail* when asked about the predicted economic impact that "It's nothing, a quirky, six-month recession would be worse than that. What it shows is that it's not going to cost us an arm and a leg. It's not going to bankrupt the country."[6] In contrast, Nancy Hughes Anthony, president of the Canadian Chamber of Commerce, said that "Canadians should be alarmed by this figure."[7]

Overall, the results from the macroeconomic analysis suggest that businesses must invest an additional $10-$16 billion (1997 prices) for energy efficiency and to directly reduce GHG emissions.[8] Many businesses – construction, companies that manufacture and supply building materials, appliance and electronic equipment makers, and vehicle manufacturers – would benefit from the increased spending that results from actions that target their respective products. Spin-offs from this increased activity, such as higher earnings, indirectly stimulate other businesses. However, after this initial positive effect, negative impacts emerge. Investment in the reduction of GHG emissions generates a benefit, in part, for which there is no market-based return, only the reduction in GHG emissions.[9] Productivity is therefore reduced as measured by rising input costs of labour and capital relative to output (*total factor productivity*). By 2018 total factor productivity is reduced by 2.3%. Companies have to raise prices to cover rising costs, and these higher prices reduce consumer incomes.

Competitiveness impacts on businesses are variable across regions and sectors because energy efficiency and fuel switching are cheaper in some sectors than in others. Table 4.4 indicates impacts to GDP, productivity, and unit costs for major types of firms. Sector and regional impacts are discussed in the next two chapters.

The difficulty in estimating macroeconomic impacts cannot be overstated. The results presented here have a high degree of uncertainty, especially at higher GHG permit trading prices. Also, it is important to remember that the microeconomic and macroeconomic models were run sequentially rather than simultaneously. This means that CIMS achieved the Kyoto target with direct actions only and was not able to rely on reductions in the output of some industrial sectors (other than the energy sector). In reality, both microeconomic and macroeconomic responses to policy would occur together and possibly reduce the permit-trading price needed to achieve the Kyoto Protocol.

Table 4.4

Competitiveness impacts in 2010, national target (% change from reference case)

Firm category	GDP (millions $1986)	Unit cost changes (GDP deflators)	Productivity (total factor productivity)
Resource-based goods	-3.7	4.1	-5.0
Durable and investment goods	-4.9	6.4	-3.1
Business-related services	-0.9	4.0	-2.0
Consumer goods and services	-1.6	7.3	-0.2
Government and social services	-0.1	6.7	-0.1

Source: Informetrica, *Macroeconomic Impacts of GHG Reduction Options: National and Provincial Effects* (Ottawa: Informetrica, 2000), Tables 155, 157, and 158.

Impact on Family Income and Expenditures

Family income and expenditures depend on factors such as employment levels, wage rates, price levels, and government spending on education and health. To gain a sense of how families and individuals may be affected by GHG emission reductions, it is useful to consider *disposable income per household*. It indicates how meeting the target will affect the income available to households to either save or spend. Initially, disposable income is positively affected but diminishes to negative changes as the target date approaches. In the long term, a reduction in disposable income per household of approximately 4% is foreseen (Table 4.5). This pattern follows the one described for GDP impacts – investment initially stimulates the economy, improving wage rates, but later declining productivity and higher factor costs cause employment and real wage rates to be reduced.

The changes that consumers will see in their utility bills depend on the type of policy. The financial signal provided by permit trading or GHG taxes increases the delivered price of energy, thus influencing households to use energy more efficiently and to switch to fuels that contain less carbon. Based on the cap and tradable permit modelling results, the following changes to consumer energy prices would occur under the national target.

Electricity: Electricity price increases depend on the region and the pricing strategy. We assume that final price in each province will still be based on average cost, which is the current model for regulated electricity prices in most provinces. If provinces allow electricity pricing based on marginal costs, then price increases could be similar from one province to another, because in most provinces a combined-cycle gas turbine is considered to be the marginal generation investment. With average cost pricing, price increases are significantly higher for provinces such as Alberta and Saskatchewan, which have higher shares of coal generation, and negligible for hydroelectric provinces such as Quebec, British Columbia, and Manitoba. See Chapter 5 for provincial details (Figure 5.8); price increases range from just 2% to 84%. The price increases represent both the permit costs and the increased costs of production that are passed on in electricity prices. They also reflect the effect of market changes to demand. In terms of what consumers could

Table 4.5

Impacts to households, national target (% change from reference case)

Impact	2010	2018
Employment	-1.45	-1.09
Disposable real income per household	-2.44	-3.67

Source: Informetrica, *Macroeconomic Impacts of GHG Reduction Options: National and Provincial Effects* (Ottawa: Informetrica, 2000), Table 154.

face, an electricity price of 6.5 to 9.0¢/kWh would rise to between 6.7 and 14.9¢/kWh depending on the region. Electricity costs to heat a typical 1,200-square-foot, single-storey home with a heated basement in Toronto would rise to about $1,800 a year from $1,400.[10]

Natural Gas: Prices could increase by 40% to 90% if the permit price is passed on to consumers in the delivered price of natural gas. Residential natural gas prices of $10/GJ in Ontario would rise to $16/GJ. Natural gas costs to heat and provide water heating for a typical 1,200-square-foot, single-storey home with a heated basement in Toronto would rise to $1,900 a year from $1,200.[11]

Gasoline: After-tax prices could increase by about 50% if the full permit cost is passed on to consumers in the delivered price of gasoline. Thus, a gasoline price of 73¢/litre would rise to $1.10/litre. For a typical driver, this translates into a rise in annual fuel costs from $1,500 to $2,300.[12]

Co-Benefits that May Lead to a Higher Quality of Life

Taking action to reduce GHG emissions has been advocated by many as a way to achieve other societal objectives. Awareness of how actions affect other aspects of quality of life (social, environmental) formed an important element in the analysis, though this element was mainly assessed qualitatively by Issue Tables in their Options Papers. Many actions reduce emissions by using resources more efficiently and thus reduce the amount of waste with which ecosystems must cope. Municipalities reduce energy consumption by diverting waste from landfill to recycling. Water conservation saves both energy and water. Actions that switch energy consumption away from carbon-intensive fuels reduce the amount of local air pollutants released into the environment. Some actions contribute in other ways to environmental objectives. For example, improved soil stabilization and water quality result from afforestation and urban forestry actions.

While pursuing GHG emission reductions usually leads to significant benefits that can improve quality of life, this is not always the case. For example, increasing the airtightness of buildings to improve energy efficiency can cause indoor air quality to worsen in the absence of efforts to mitigate this impact. Also, equipment adopted to meet other environmental needs can increase CO_2 emissions. For example, meeting increasingly stringent sulphur content requirements for gasoline requires refineries to employ equipment that increases CO_2 emissions.

Lifestyle Impacts

Lifestyle impacts are, in part, already implied by the above sections. Family well-being will be relatively unaffected in terms of the ability to maintain material standards of living. However, targeted reduction programs or increased energy prices will drive people to do some things differently. Many of these actions are described in the transportation and residential sections in the following chapter. Changes in lifestyle are assumed to contribute only minimally to the GHG reductions represented here. For instance, people will not be living with lower indoor temperatures or less lighting.

The most significant lifestyle changes will occur in transportation. Two key actions in this area are increased transit use through improved public transportation and the reduction of travel speeds on highways and freeways. By expanding transport infrastructure and improving service, an estimated 2% more people will take transit. Travelling at posted highway speed limits would reduce emissions by 4.2 Mt CO_2e (national target). While some argue that such changes bring about an improvement in lifestyle, others see them as a cost, a loss of personal welfare. Estimation of these costs, losses in consumers' value, is included in our cost estimates.

The Results in Context

In this section we relate some of the findings of CIMS's integrated analysis with other key studies. Given that the approaches and aims of these studies are quite different, we make no attempt at disaggregated comparison. Instead, we look at how our results compare with the key cost conclusions of those studies.

MARKAL

As noted in Chapter 3, MARKAL, an optimization microeconomic model, also tested emission reductions in Canada for the NCCP costing study.[13] Table 4.6 compares the findings of this study with those of CIMS. Not surprisingly, MARKAL's analysis came to the same conclusions about strategies to reach the Kyoto target; it is less costly to reduce emissions under a national target. However, there are significant differences in the magnitude of the long-term impact on GDP and in the permit price necessary to meet the Kyoto target.

MARKAL determines emission reduction actions strictly on the basis of financial costs. Thus, a lower financial signal of \$50/t CO_2e is sufficient to motivate change in equipment acquisition behaviour. In contrast, because

Table 4.6

Comparison of permit-trading price and GDP impact: CIMS, MARKAL

Model	Permit trading price (\$/t CO_2e)	Direct total costs (billions \$1995)	GDP impact[a]
MARKAL	\$49	20	-0.5%
CIMS	\$120	45	-2.1%

a "GDP impact" refers to the long-term impact in 2018 (annual % change from reference case) reported by Informetrica.

consumers and producers in CIMS make technology choices based on preferences and attitudes to risk, in addition to financial cost, they require a higher cost incentive to change behaviour. Thus, the total cost estimate of the national target for MARKAL is $20 billion compared with the $45 billion for CIMS, and the GDP impact with MARKAL is less than 1%.

National Energy Modeling System (NEMS), US Department of Energy Study

The US Department of Energy used the NEMS model to analyze the impacts of the Kyoto Protocol on US energy markets and the economy.[14] The methodology of NEMS is similar to that of CIMS. A financial signal is attached to energy, simulating the price effect of tradable GHG permits or a GHG tax, so that emission reductions are made across the economy in a least-cost manner.

The permit price required to reduce US energy-related carbon emissions to a target of 3% below 1990 levels in the period between 2008 and 2012 is $294 per tonne of carbon in 2010 in 1996 $US (and $348 per tonne of carbon to meet a target of 7% below 1990 levels). The result for 3% below 1990 levels for the United States is most directly comparable to Canada's Kyoto target, because at these targets both studies exclude the effect of forestry and agriculture sinks. (When the same approach is applied to Canada, reductions of only 4.3% from 1990 levels are required.) When the CIMS permit-trading price of $120 Cdn/t CO_2e for this target is converted to US currency and to carbon units, it equals about $300 US per tonne of carbon, very close to the $294 of the NEMS study.

The similarity of the NEMS and CIMS results is not surprising when one considers the parallels between the studies, models, and their subject matter. Both the United States and Canada have similar targets, and the growth trend of emissions in both countries in the decade since 1990 was similar. Also, both economies have similar energy intensities, industry, and geography, although Canada is more focused on resource extraction and processing than the United States. Both studies involve similar, economy-wide policy instruments. Most important, though, the two models represent technologies and the preferences of businesses and consumers in similar ways. Thus, the estimated tax level or emission permit-trading price is based on firm and household perceptions of costs and benefits, not just on financial costs.[15]

MS-MRT Model

The Multi-Sector Multi-Region Trade (MS-MRT) model of Charles River Associates has been used to estimate the costs of achieving Kyoto targets for industrialized countries when they acted alone or traded among themselves.[16] The estimates for Canada acting on its own are relevant for this study. In a recent application of the MS-MRT model, it was estimated that Canada would need a tax level of \$283/t carbon (\$US 1995), equivalent to \$115/t CO_2 (\$Cdn 1995). This results in a reduction of 2.14% of GDP in 2010, again very close to the results of our study.

MS-MRT is a computable general equilibrium (CGE) model, one that has both micro- and macroeconomic representation. At the microeconomic level, the response to technologies and prices by firms and households is represented via aggregate production functions for firms and aggregate utility or consumer functions for households. At a macroeconomic level, the model links changes in the output costs of goods and services, and consumer responsiveness to these costs, to aggregate macroeconomic balances, such as total investment, interest rates, government expenditures, sector adjustments, and trade. Assumptions about the ease with which energy can be substituted with other aggregate inputs, such as labour, capital, and materials, and the ease of substitution between energy forms are important to how it models GHG reduction actions. Important too is its representation of the natural evolution (not price induced) of technology toward or away from energy. The estimated cost of reaching the target of \$115/t CO_2 suggests that these assumptions used in the MS-MRT model are, in combination, similar to the implicit values in the technology and consumer preference information used in CIMS in this study.

Uncertainty Issues and Sensitivity Analysis

In Chapter 2 we listed many of the key assumptions that go into the estimation of costs. Some of these assumptions are contentious, and thus uncertain for the decision maker, because there is disagreement on the definition of key concepts, indeed the concept of cost itself. Other assumptions are associated with a great deal of forecasting uncertainty for developments internal and external to Canada.

Sensitivity analysis is an exercise in which one tests for the importance to the modelling results of changes in key parameters, data inputs, or external assumptions. We have gathered the results from sensitivity analyses conducted by us and other researchers in order to pinpoint uncertainties of particular concern.[17] In this section we follow the general organization of Chapter 2 in order to cover all issues related to cost estimation.

Baseline Assumptions

We did not test an alternative reference case in this study. This does not mean that we would not value the lessons from such an exercise. Indeed, we recommend this approach for future work. However, the heavy data handling and model operation related to the high number of actions proposed by the Issue Tables made it difficult to envision this additional time- and modelling-intensive step. In any case, it is our judgment that the marginal costs of meeting Canada's reduction target will not change substantially if future population and economic output by 2010 are no more than 5% higher or lower than their forecasted levels. Total costs will of course change with the different growth and emission levels, but within these ranges the change in costs is likely to be proportional to the change in output. Thus, for example, if a new reference case indicates that Canada's emission reduction in 2010 must be 200 Mt CO_2e instead of the 180 Mt CO_2e assumed here, the model results from this study can provide an estimate of the marginal and total costs associated with this bigger target.

Assumptions about the Definition of Costs

The debate over the definition of cost is especially important with consumption goods – personal vehicles, appliances, and houses – where estimates of consumers' value are important to cost estimation. We tested this concern in the transportation sector. Not all actions identified by the Transportation Issue Table included consumers' value loss. When we incorporated estimates for it, we found that, for the sector to achieve the same amount of emission reductions as it did under the national target (58 Mt CO_2e), the marginal cost increases from $120/t CO_2e to $210/t CO_2e. Total costs in the transportation sector rise from $19.5 billion to $27.9 billion. When the entire model is run for Path 4 with this adjustment, the permit-trading price to achieve the Kyoto Protocol increases to $150. Also, the contribution of transportation falls from 24% to 17%, while that of electricity rises from 43% to 47%, with other sectors staying about the same.[18]

If these estimates of changes in consumers' value are accurate, then they suggest that consumer resistance to GHG emission reduction policies may be critical to cost estimates. Policies may have better prospects when they do not require too dramatic a change in behaviour. For example, a shift from gasoline cars to ethanol or even fuel cell cars may be easier than a shift from the automobile to public transit. While a strict financial analysis might show these two alternatives to be of equal cost, an analysis that includes consumers' preferences could show the relative costs to be very different. At the same time, there is some hope that carefully designed policies may contribute to changes in consumers' awareness of their opportunities and that these changes themselves can lead to different preferences and lower cost estimates. We discuss policy design in Chapter 7.

Assumptions about Preference Change

Preferences may or may not adjust in response to future policies and public concerns. As a product becomes better known in the market, the willingness of consumers to accept it undergoes a transformation in which many intangible concerns can decrease significantly. This implies that the losses in consumers' value associated with adopting this technology decrease. Different tests of plausible market shifts suggest that the costs of

GHG emission reduction in the transportation sector could fall significantly if policy succeeds in motivating a market transformation.

Our research group recently probed the potential effect on cost estimates of preference shifting in the transportation sector, notably in terms of the willingness of consumers to accept alternatives to the gasoline-fuelled internal combustion engine.[19] The study tested different assumptions about how much preferences would change after a vehicle emission standard had been implemented. (The vehicle emission standard policy is described in Chapter 7.) If low-emission and alternative-fuel vehicles are eventually seen as near-perfect substitutes to the internal combustion engine, then the marginal cost for the transportation sector to achieve its previous share of the national target falls to $100/t CO_2e, which implies that the marginal cost for reaching the national target would also fall. In the time frame to 2010, ethanol vehicles were the strongest-performing alternative, in part because the tight time frame of the Kyoto Protocol hindered the ability to achieve a substantial commercialization and diffusion of fuel cell vehicles.

Assumptions about Technological Change

As noted in Chapter 2, there is a great deal of uncertainty about the pace and tendency of technological change in the race to decarbonate energy systems. Salient options include nuclear, energy efficiency, renewables, and CO_2 capture and sequestration from fossil fuels. The cost and the potential magnitude of this last process are particularly uncertain. We report in Table 4.7 the results of a sensitivity analysis in which the base assumptions about the availability of CO_2 capture and sequestration are altered from being 50% available to not at all available.[20]

As one can see from these estimates, changing the assumptions about this technology does not have a large effect on the total costs. However, we caution that there are studies suggesting that, as one moves beyond the Kyoto time frame of 2010, the development of clean fossil fuels could have large effects on the costs of GHG emission reduction.[21]

Table 4.7

Sensitivity of the availability of CO_2 capture and sequestration to cost

Application of action	Share of electricity emission reductions	Displaced reductions (Mt CO_2e)	Permit-trading price ($/t CO_2e)	Total cost (billions $1995)
Full	35%	0	120	44.7
Half potential	17.5%	13.5	127	45.0
No potential	0%	27	135	47.7

Note: Based on sensitivity results from MARKAL's NCCP analysis, adjusted to reflect our action penetration rates and costs.

Assumptions about Electricity System Integration

No action's costs should be estimated in isolation because cost is always influenced to some extent by other actions that are simultaneously undertaken. However, the operation of an integrated model requires its own set of assumptions about how integration in the energy system will unfold in the future. In most cases there is not a great deal of uncertainty since the interconnections of the economy are well known and stable. But there are cases in which they are less certain in the future.

One uncertainty that stands out is the future interconnection of provincial grids and electricity markets. In the past provincial electricity markets were operated as enclaves with minimal interconnection. However, wholesale electricity markets throughout North America are moving toward greater integration and more open access. Will this trend continue? Will it allow for a dramatic increase in electricity trade between provinces, much of it motivated by GHG policy? Will this trade in turn reduce the costs of GHG emission reduction? Table 4.8 shows the results of a sensitivity analysis in which the incremental effect on costs is analyzed for two situations, one allowing for greater interprovincial electricity connection and trade and one assuming that interconnection will not change a great deal from its current levels.[22]

Allowing for greater interconnection of provincial grids and greater interprovincial electricity trade does indeed decrease the costs of GHG emission reduction. However, the cost decrease is not great by itself. Total costs decrease from $44.7 to $41.7 billion. This relatively small decrease is explained in part by the assumption that the incremental electricity generation resource in most provinces is natural gas. With existing hydropower facilities fully utilized, increased trade opportunities between provinces do not lead to dramatic decreases in the thermal production of electricity. However, when the time frame is extended beyond 2010, and major hydroelectric projects and transmission upgrades are assumed to be possible, the cost effectiveness of interprovincial trade increases dramatically. (Of course, this scenario requires the additional assumption that Canadians will support large hydroelectric projects and substantial expansion of high-voltage transmission.)

In one of our sensitivity analyses, with admittedly extreme assumptions about large hydropower and transmission development, we found that the

Table 4.8

Sensitivity analysis of interprovincial electricity trade

	Mt CO_2e reduced	Marginal cost ($/t CO_2e)	Total cost (billions $1995)
No extra trade	0	120	44.7
Extra trade	3	118	41.7

electricity sector could achieve its contribution to the national target at a permit-trading price of \$60/t CO_2e.[23] If the full integrated model were tested, then the result would be a nationwide permit price of about \$100/t CO_2e.

Assumptions about Macroeconomic Costs

The macroeconomic or indirect costs of policies on GHG emission reduction are a source of uncertainty in that it is difficult to know the extent to which one sector of the country or region will be impacted, given all the factors involved in determining the competitive prospects for Canadian firms. In this study we did not conduct a sensitivity analysis of aggregate macroeconomic effects.[24] One reason is that the aggregate macroeconomic effects were not that substantial for the target that was tested, although some industry sectors could be significantly impacted even at this target level.

Assumptions about Other Benefits from GHG Emission Reduction Actions

GHG emission policy is accompanied by significant secondary impacts. In Chapter 2 we noted that these impacts can be monetized and included in a cost-effectiveness analysis or characterized within a multiple account framework in either monetary terms or physical terms. Table 4.9 explores the ways in which the air quality impacts (as quantified by the Environment and Health Impacts Subgroup of the Analysis and Modelling Group) of our

Table 4.9

Representation of air quality impacts (national target)

Criteria air contaminant (CAC) emissions (% change 2010)		Impact on ambient air quality in 2010 (maximum reduction in concentration)		Avoided health event (2010)[a]	
PM2.5	-8%	SO_2	2.38 ug/m^3	Deaths	92
SO_x	-14%	Ozone	0.5 ppb	Chronic bronchitis	320
NO_x	-14%	SO_4	2.38 ug/m^3	Respiratory hospital admissions	58
VOC	-10%			Asthma symptom days	140,000

a Mean in range.

Source: Environmental Health and Impacts (EHI) Subgroup, Analysis and Modelling Group, *The Environmental and Health Co-Benefits of Action to Mitigate Climate Change* (Ottawa: Analysis and Modelling Group, NCCP, 2000), Tables 3-21, 3-13, 3-26a. The EHI subgroup's results are for CIMS's Path 2; Path 4 energy changes are almost identical.

modelled policies can be represented. The column on the far left shows the most straightforward depiction, which takes our energy results and through factor relationships transforms them into changes in the production of key criteria air contaminants (CAC). This is the most methodologically straightforward step, though it is not without its uncertainties since (unlike GHG emissions) the level of CAC emissions is dependent on many factors, such as air temperature and combustion technology, in addition to fuel type. Estimates of ambient air quality impacts and avoided health events involve additional methodological complexity and uncertainty. A dollar figure can be put on the avoided health effects: the Environment and Health Impacts Subgroup estimated air quality benefits to be $2.68 billion. However, because there is considerable uncertainty in monetization, it is wise to consider this estimate separately from our results.

Assumptions about US Policy and GHG Reduction Agreement Parameters

Most of the simulations for this study are undertaken with the assumption that the Kyoto agreement provides the national target around which to focus our analysis. However, even with the Kyoto agreement, there is uncertainty about what might be required in terms of domestic GHG emissions reductions. Two key uncertainties are the definitions of forestry and agricultural sinks. Another is the amount of national commitment that might be achieved by purchasing reductions from another country. Hence, analysts have emphasized the importance of generating cost curves of GHG emission reduction. CIMS is well suited for this exercise because a new set of results is generated every time a new permit-trading price is tested in the model.

Figures 4.7 and 4.8 show curves that depict the permit-trading prices required to elicit different levels of sectoral and total GHG emission reduction in the Canadian economy in the 2010 time frame. These curves were generated by simulations of CIMS, and they provide key information for negotiations on the Kyoto agreement and for future agreements and targets. In particular, they show what different targets would cost Canada and how much Canada would purchase externally if some of its commitment could be acquired through an international permit-trading system.

Insights can be gained from Figure 4.7 on the uncertainties of sinks and external acquisitions. If Canada were not allowed to count its sinks in the way that it has estimated them, then the amount of domestic reductions would increase, meaning that the dashed vertical line in Figure 4.7 would move to the right and that the marginal cost would increase. Additional reductions would be drawn from different sectors according to the shape of their marginal cost curves in Figure 4.8, and according to the equi-marginal principle. Conversely, buying emission reductions from abroad would reduce

Figure 4.7

National GHG emission abatement cost curve, calculated from CIMS results (2010)

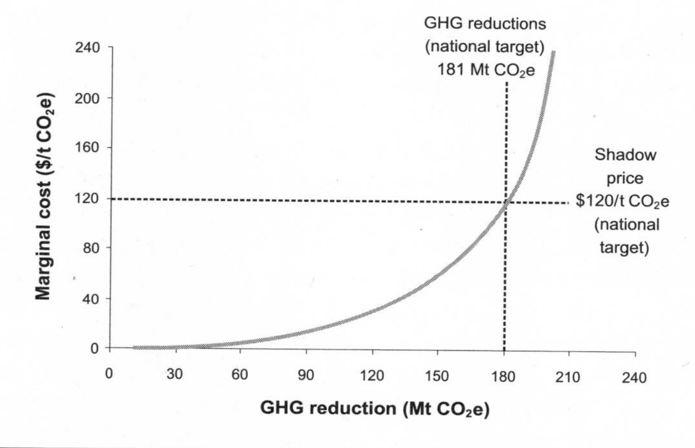

Figure 4.8

Sector-specific GHG emission abatement cost curves, calculated from CIMS results (2010)

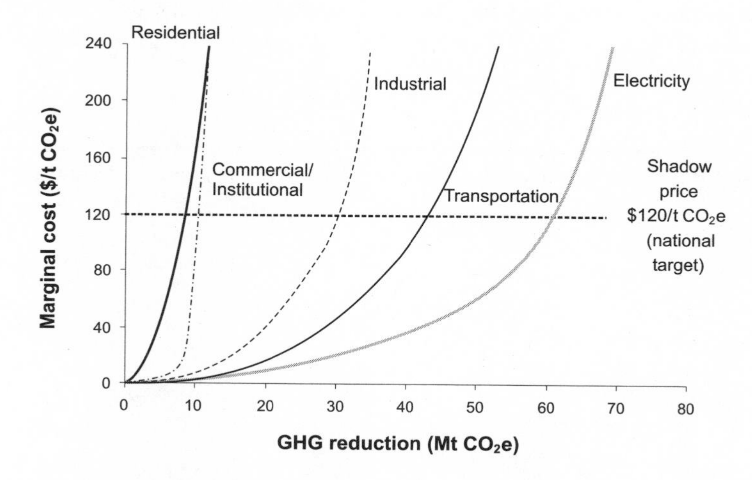

the amount of domestic reductions that Canada must make, freeing sectors from their commitments, again according to the equi-marginal principle.

We explored in more detail the cost implications of using international trade to meet Canada's emission reduction target. It is generally assumed that efforts to broaden the global involvement to reduce GHG emissions will provide lower-cost opportunities for Canada. We tested two cases in which Canada is able to purchase some of its emission reductions from other countries at lower costs: one at a permit price of \$22/t CO_2e and one at \$55/t CO_2e (these are average prices over the modelling period).[25]

When Canada acts alone, 181 Mt CO_2e are reduced domestically. Actions are employed according to cost, so that relatively cheap reductions give way to more expensive actions. The 181st tonne of GHG reduced has a cost of \$120. Under the sensitivity scenarios, Canada is not limited to its own reduction opportunities. International opportunities are expressed through an international permit-trading system whose permit prices are dictated by the marginal cost of GHG reductions internationally – those who have higher costs to reduce emissions domestically will buy permits, and those with lower costs will sell them. When Canada has access to this mechanism, domestic reductions that are cheaper than the international permit prices will still be undertaken, but those that exceed the prices will be abandoned in favour of buying permits. As shown in Figure 4.9, the international permit prices of \$22/t CO_2e and \$55/t CO_2e are substantially lower than the permit price of \$120/t CO_2e when Canada acts alone, and domestic reductions are reduced by 51 Mt CO_2e and 30 Mt CO_2e respectively. This suggests that the bulk of reductions, approximately 130 Mt CO_2e of the 181 Mt CO_2e

Figure 4.9

Marginal costs and emission reductions in 2010 with international permit trading

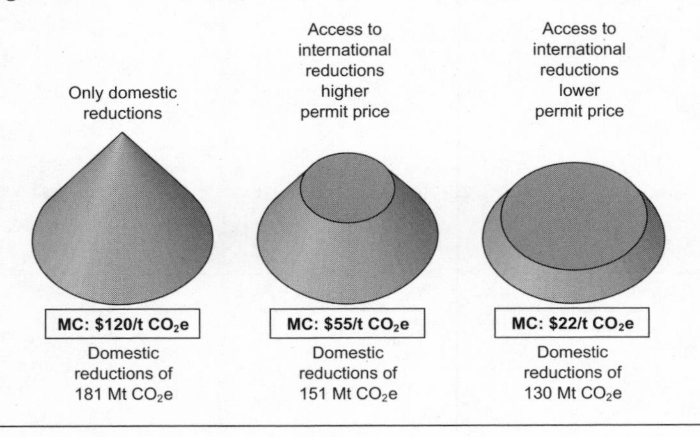

Figure 4.10

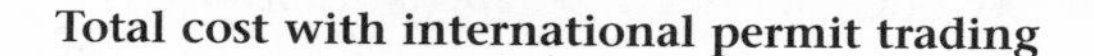

Total cost with international permit trading

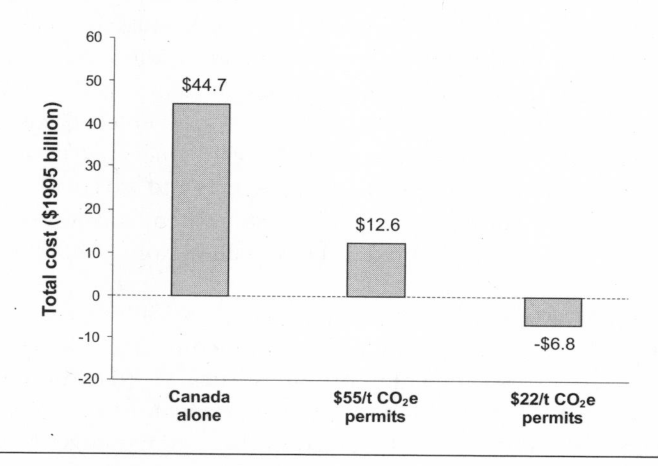

to meet the Kyoto target domestically, occur from actions that cost $22/t CO_2e or lower; this is confirmed by the shape of the cost curve in Figure 4.7. So while the marginal cost for domestic reductions is $120/t CO_2e, most actions cost significantly less.

If Canada can purchase international permits, then total costs are reduced significantly. Figure 4.10 shows this reduction graphically for the national target. The more that Canada can access low-cost international emission reduction opportunities, the more higher-cost domestic measures can be avoided. In fact, total costs become negative (benefits) because high-cost measures, once removed, leave low-cost and profitable measures in place.[26]

Conclusion

In this chapter we have explored how our results relate to the many complex issues that surround cost estimation and what they imply should Canada's commitments change. While our results focus on one particular level of emissions abatement, we emphasize that understanding how the costs might differ for alternative levels of emissions abatement is critical for future domestic goal setting and international negotiations.

The debate over cost definitions is especially important to cost estimates of actions affecting final consumption. For example, a more complete estimate of the willingness of automobile drivers to reduce vehicle use might mean that the permit-trading price for achieving the Kyoto target will be higher, perhaps $150/t CO_2e, compared with the $120/t CO_2e reported in our results. Also important is how technologies, such as CO_2 capture and sequestration, will evolve, particularly as we move beyond the Kyoto time frame of 2010. We return to these uncertainties in Chapters 7 and 8.

Achieving enough GHG reductions such that Canada's emissions are 6% below what they were in 1990 by 2008-12 will influence the way that our communities and economy develop and ultimately affect the standard of living. This and other economic impacts should inform policy design, an issue that we examine in Chapter 7. Although Canadian economic growth and the competitiveness of businesses will likely be negative with a Kyoto target, this impact varies across sectors and regions and, taken as a whole, is relatively moderate. This is particularly the case when emission reductions are not fixed for each sector, with direct costs falling from $61 billion to $45 billion.

A look at other elements that contribute to quality of life, including family income, job prospects, community infrastructure and services, and air quality and other environmental attributes, suggests that the impact of meeting the target will not be particularly onerous for individuals. However, the tradable permit policy assumed in the national target simulation will have the same impact as a tax on prices, resulting in a considerable increase in energy costs faced by consumers. Also, making reductions involves changes in people's preferences for certain technologies, goods, and services; some of these changes will not occur easily.

But where are our results situated within the larger debate between top-down and bottom-up modellers? Like bottom-up models, CIMS is technologically explicit. This feature should allow it to show a greater potential for technological change, which would lead to lower long-run costs of GHG abatement. However, the Kyoto deadline of 2010 is too tight for much of this advantage to materialize. Unlike a standard bottom-up approach, CIMS includes consumer inertia in its portrayal of market dynamics and includes consumers' value in its assessment of costs. This explains why CIMS does not generally show GHG abatement to be a profitable or a low-cost proposition. There is a cost to the economy and a significant impact on energy prices to achieve the Kyoto target. In this sense CIMS's results may be seen as closer to a top-down model. However, CIMS also shows how declines in equipment costs over time (perhaps with focused research and commercialization efforts for innovative technologies) and changes in consumer preferences (with greater marketing efforts) can lead to lower long-run costs of GHG abatement.

Is the cost of the Kyoto target high or low? Perhaps a valid answer is that it is both high and low depending on the perspective, which may indeed explain the divergent estimates of the top-down and bottom-up approaches to GHG abatement costing.

The cost is high if you consider that certain preferences of businesses and consumers for specific technologies and ways of doing things will be difficult to change, which is why energy prices need to increase so much in order to trigger a significant reduction in emissions (which is consistent with historical data on the small response of consumers to gasoline price changes). Given the intolerance of consumers, who are also voters, for energy price increases of any magnitude, the political reaction against price increases will likely be in full force long before those increases can attain the levels necessary to enact the changes required by a GHG target such as the Kyoto Protocol. From the perspective of the politician, therefore, the costs of GHG abatement are indeed high.

However, the cost is low if you consider that energy expenditures comprise less than 5% of the budgets of most consumers (and all but a few energy-intensive industries), so that even a 50% increase in the cost of energy would entail at most a 2.5% decline in income. And some of this decline is a short-term phenomenon. The higher prices would induce investments in more energy-efficient equipment and fuel switching to lower-cost energy forms such that over time the total budgetary effect is reduced. Indeed, some of these technologies will be lower cost on a strict financial basis, and over time even consumers' value losses will be reduced as manufacturers of more energy-efficient products learn to provide better substitutes for conventional technologies (thus, one might expect that compact fluorescent light bulbs will become indistinguishable in quality from incandescent light bulbs). Even the costs to energy-intensive industries may not be too great, especially if our trading partners enact similar policies to meet similar targets and if domestic policy is designed to minimize the competitive disadvantage for domestic industry (as some Scandinavian countries have done for their energy-intensive industries when enacting GHG taxes). Also, budgetary impacts of higher permit prices can be reduced by grandfathering and some auctioning (with auction revenue recycled) of permits. These methods ensure that any cost impact results only from incremental investments to reduce emissions, not from the policy mechanism itself.

That the costs may be both large and small depending on the perspective is a critical insight in our view. Indeed, we believe that a key value of our costing methodology and disaggregated results will be to help policy makers design policies that minimize perceived short-run cost impacts while maximizing the signals needed for long-run technological innovation and preference shifting. This transitional strategy is essential if society is to overcome the current impasse caused by perceived high costs. We return to this consequence of our method and results in our discussion of GHG abatement policy in Chapters 7 and 8. Some readers may wish to skip to these chapters now. For those readers who are interested in understanding more about sector-specific actions, regional impacts, and sectoral impacts, Chapters 5 and 6 provide details.

5
Sectoral Estimates

So far we have given only a glimpse of the changes in certain sectors that contribute to emission reductions. Do the results suggest that Canadians will be walking or cycling to work, or will they be driving super-efficient or alternative-fuel vehicles? Does turning off the hot water tap contribute substantially to meeting the target, or does that occur only when big industry alters key processes? In addition, the previous chapter revealed little about how sectors could be impacted economically by making GHG emission reductions. The impact on economic growth may be relatively low, at 3% of cumulative GDP in 2010, but what does this mean for economic output in the oil and gas sector or for pulp and paper plants? In this chapter we present results by sector and show more clearly how the sectors contribute to and are affected by emission reductions. We focus on two contrasting paths, the sector target (Path 1) and the national target (Path 4). We look at the specific actions that take place when a policy is simulated. While we do not focus on policy implications, this chapter on sectoral impacts informs the policy discussion of Chapter 7. Regional implications are discussed in Chapter 6.

We first review the relative contributions of sectors to meeting the reduction target as well as the distribution of costs across sectors. In subsequent sections we delve into the results for specific sectors: residential, commercial/institutional, transportation, electricity, industry, and other (agriculture and afforestation). We present information about the sector that is pertinent to emission reduction potentials and impacts. We then describe emission reductions and the changes that occur in the sector to reduce emissions. Sector costs and related impacts are also described.

Sector Comparison

As noted in Chapter 4, there are substantial differences in cost and emission reductions when reductions are made according to sector-specific targets than when they are made in response to a national target. These differences are based directly on the dynamics and structures of the different sectors in

Figure 5.1

Sectoral contribution to emission reductions in 2010 (sector and national targets)

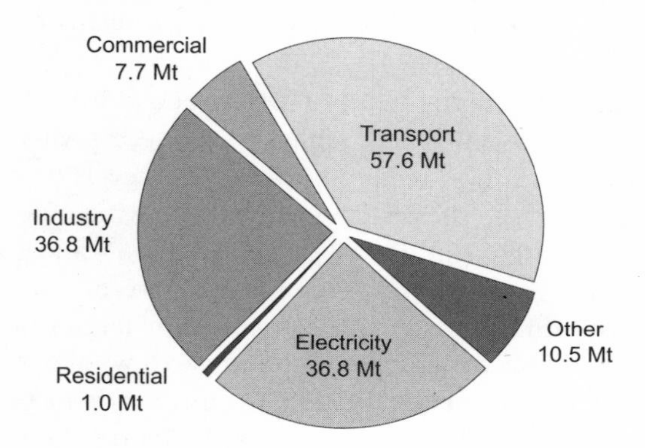

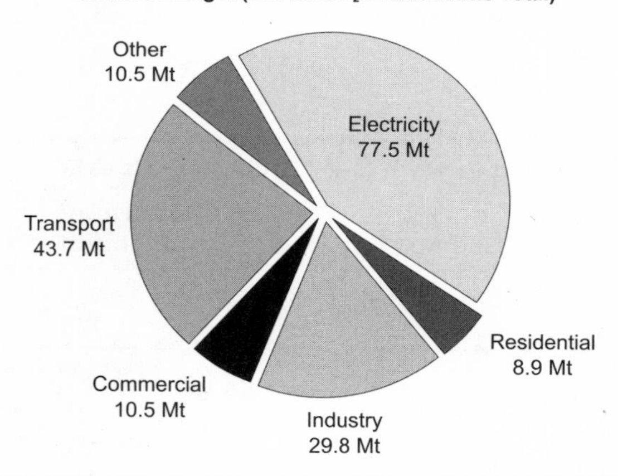

the Canadian economy. Figure 5.1 reveals the different contributions by sector to emission reductions under the sector and national targets.

In all runs the electricity, industry, and transportation sectors contribute most to emission reductions. Electricity, residential, and commercial/institutional sectors contribute more to reductions when sector targets are transformed into a national target, while the industry and transportation sectors contribute less. The contribution of electricity to total emission reductions nearly doubles from 24% to 43%, residential contributions increase from 1% to 5%, and commercial/institutional contributions increase marginally from 5% to 6%. Industry decreases its contribution from 24% of emission reduction to 16%, and transportation decreases its from 38% to 24%. The changes occur because the integrated model re-allocates actions across sectors in order to minimize total costs. Sectors are not constrained by a sector target, so that sectors with cost-effective actions beyond the sector target contribute more to emission reductions, and sectors that have to draw on relatively expensive actions to meet a sector target are freed from these actions.

Given that emissions are the result of diverse activities and technologies, we would not expect costs to be the same for each sector when sectors must meet an identical percentage reduction from 1990 levels (the sector target simulation). This is indeed the case, as can be seen by the different marginal costs for the sector target simulation in Table 5.1. Oil- and gas-processing emissions (methane for both and carbon dioxide for oil) and landfills are not subject to the financial signal. How emission targets are distributed (across sectors or regions or across the economy as a whole) makes a fundamental difference to cost incidence – where emission reductions are made and at what costs. In addition to differences based on the economic and technological makeup of a sector, marginal costs are also based on the amount of GHG emissions that each sector needs to reduce to meet an

Table 5.1

Marginal cost required to reach target, by sector (\$/t CO_2e)

Target	Electricity	Industry	Residential	Commercial/ institutional	Transportation
Sector	30	300[a]	0	10	50
National	120	120[b]	120	120[c]	120

a Did not attain target.
b Except for oil and gas processing (CH_4), oil processing (CO_2).
c Except for landfills.

Table 5.2

Emission reductions required by sector to meet targets (Mt CO_2e) under the sector target modelling simulation

Sector	Actual emissions 1990	Target in 2010	Reference case 2010	Must reduce emissions in 2010 by
Electricity	96	90	129	30%
Industry	199	187	255	27%
Residential	49	46	47	2%
Commercial/Institutional	47	44	52	15%
Transportation	147	138	198	30%

identical percentage reduction. This is based on a sector's contribution to total Canadian GHG emissions and on how those emissions are projected to grow in the reference case. Table 5.2 shows the quantity of emissions that each sector needs to reduce in the simulation. The electricity, industry, and transportation sectors constituted the majority of emissions in 1990 and face substantial growth to 2010. They consequently face reductions of approximately 30% to meet a sector target, while the reductions required of the commercial/institutional and residential sectors are minor.

Figure 5.2 shows how the marginal cost of each sector changes when one moves from the sector target to a national target. The grey bars represent the marginal cost of emission abatement under the sector target. Each sector has a different marginal cost of reduction associated with the emissions reduced to meet its sector target. Under the national target simulation, \$120/t CO_2e is established as the permit-trading price across all sectors to meet the target, and this price corresponds to the country's marginal cost of emission reduction. The arrows in Figure 5.2 indicate how the marginal costs are either raised or lowered for different sectors to meet this economy-wide marginal cost. Thus, under the national target, the residential sector contributes actions available to it between \$0 and \$120/t CO_2e, and its contribution to emission reductions increases by 7.9 Mt CO_2e, 5% of total Canadian emission reductions. Similarly, the commercial/institutional sector contributes a further 2.8 Mt CO_2e to total emission reductions. The electricity sector, which has substantial low-cost abatement opportunities, responds

Figure 5.2

Relative marginal cost of GHG abatement between the sector target and the national target

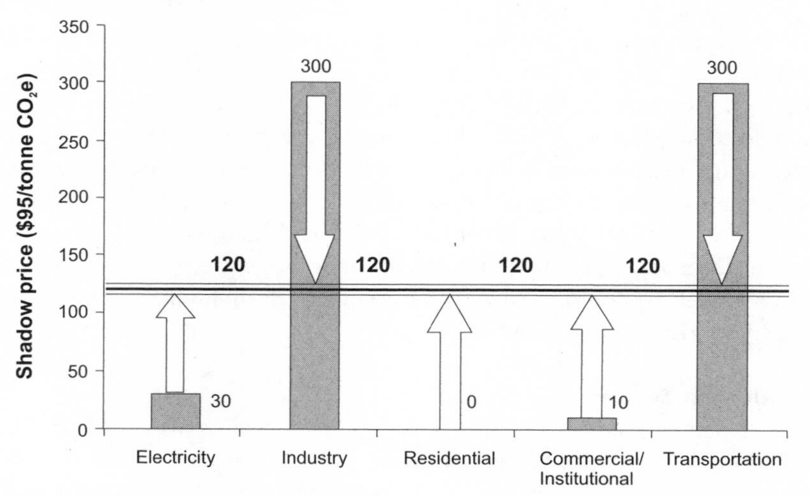

Figure 5.3

Total abatement costs by sector

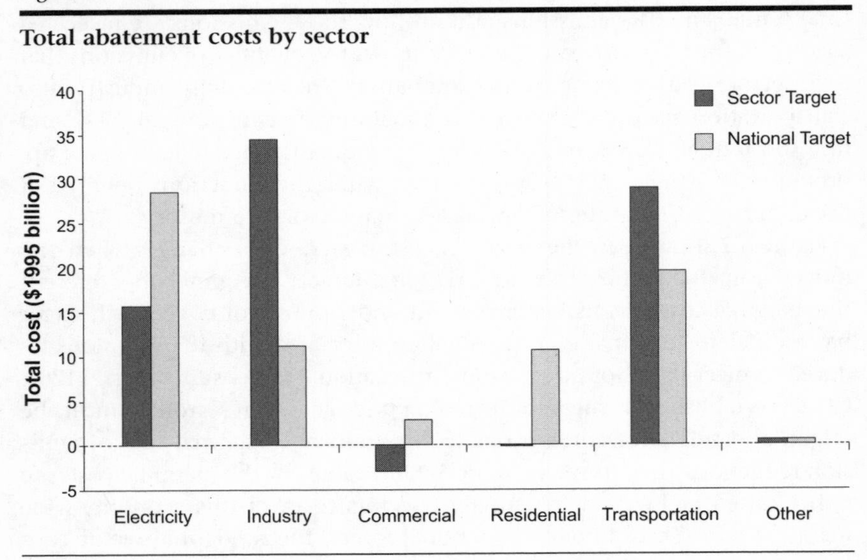

dramatically to the change in marginal cost so that approximately 40 Mt CO_2e more are reduced, increasing the electricity sector's share of reductions to nearly half of the total (43%). The other key GHG-producing sectors react to an opposite signal – transportation and industry sectors reduce their contributions by 14 and 7 Mt CO_2e respectively as they drop actions costing more than $120/t CO_2e.

Total financial costs by sector vary significantly under a sector target and a national target. These costs are shown in Figure 5.3. Note that, while we show the costs to the electricity sector in the graph, these costs are also incorporated into the costs of the demand sectors (all other sectors shown) through electricity price changes. Thus, summing up all sectors would double-count electricity costs. Total Canada-wide costs for the national target are significantly smaller than for the sector target because cost decreases in the industrial and transportation sectors exceed cost increases in the commercial/institutional and residential sectors, and therefore overall costs decrease. Costs in the commercial/institutional sector are negative (benefits) because fuel expenditure savings exceed costs. Individual sectors are discussed in considerable detail below.

Residential Sector

The residential sector is a relatively small contributor to overall direct GHG emissions; in 1998 this sector contributed 7% to national emissions.[1] This already small share is expected to decline further by 2010 in the reference case to about 6% due to a combination of factors. Electricity use is expected to grow, driven by increasing penetration and use of electrical equipment and appliances. This trend translates into increased overall GHG emissions (combined direct and indirect) in regions with significant shares of fossil fuel-based electricity generation (Alberta, Saskatchewan, Ontario, Nova Scotia, and New Brunswick) but lowers overall emissions in hydroelectric regions.

More energy-conscious building practices lead the way toward emission reductions. Regulations relating to furnaces, water heaters, and appliances are expected to increase efficiency, making a significant impact in energy consumption (and subsequent GHG emissions). These gains are offset somewhat by trends toward increasing house size and window size, exposed basement walls, larger water heaters, increased lighting, and a greater number of appliances, gas fireplaces, and in-floor radiant heating systems.

Stock turnover alone provides significant reductions in energy intensity, since new houses are more energy efficient. At the national level, more than 20% of dwellings were built before 1950, and almost 30% date from the 1950-70 period. About 50% of existing dwellings were built after 1970; few of these dwellings have ever been upgraded.[2] Given the relatively young age of most of Canada's building stock and the longevity of houses and heating systems, there is a certain amount of inertia hindering significant energy efficiency improvements over the study time period.

Many efficient technologies in today's marketplace can substantially reduce residential energy consumption and provide financial benefits. However, residential consumers have been reluctant to invest in energy efficiency, even when accompanied by ample financial benefits, suggesting some combination of nonfinancial preferences for standard technologies, perceived risks associated with switching to energy-efficient alternatives, and perhaps inadequate awareness of efficiency benefits.

Emission Reductions

Figure 5.4 and Table 5.3 show how the residential sector contributes to GHG emission reductions under sector and national targets. Reference case projections alone bring emissions close to the sector target. Thus, few additional actions are required to meet the target (only 1 Mt CO_2e). Under a national target, the residential sector is pushed significantly further, so that emissions are 22% lower than 1990 levels. Strong reductions occur because of low-cost opportunities for energy efficiency improvements. Reductions would be greater if CIMS did not take into account consumer preferences.

Source of Emission Reductions

All emission reductions in this sector are energy based. As can be seen in Figure 5.5, reductions occur both from gains in efficiency and from changes

Figure 5.4

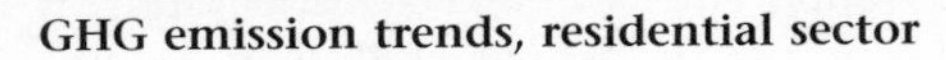

GHG emission trends, residential sector

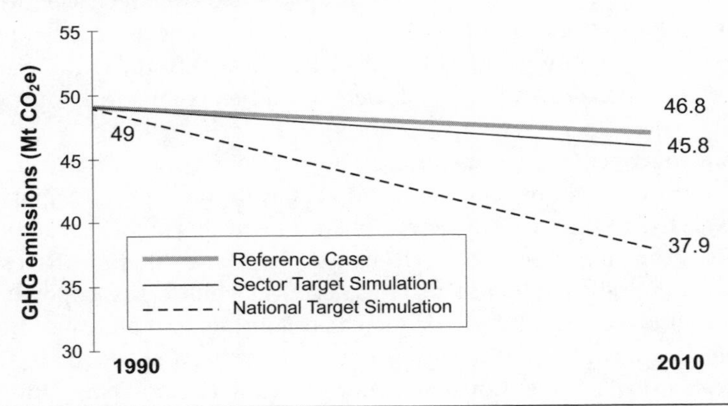

Table 5.3

GHG emission reductions, residential sector

	National target	Sector target
Change relative to reference case in 2010 (Mt CO_2e)	-8.9	-1.0
Percentage change relative to reference case in 2010	-19%	-2%
Percentage change in emissions relative to 1990 level	-23%	-6%
Contribution to national target reductions (% of total)	5%	1%

in the type of energy used. Fuel shares shift slightly between the reference case and the policy simulations, although total energy consumed drops approximately 11% under the national target and 3% under the sector target. Energy efficiency is the source of more reductions because more actions are focused on energy efficiency than on fuel switching. Also, as may be recalled, substantial fuel switching to electricity is assumed to have already occurred in the reference case. Nevertheless, switching from oil and wood to less GHG-intensive fuels and electricity is responsible for 43% of the reductions in the national target simulation.

Space and water heating are the most significant contributors to emissions in this sector: together they account for about 80% of energy use and virtually all direct emissions. It is of little surprise that most direct emission reductions occur in these end uses (87% are from space heating and the remainder from water heating). The following are key energy efficiency actions.

- Prevent heat loss by upgrading the *building shell* (altering insulation, windows, and doors) and by improving furnace efficiency. New furnaces have

Figure 5.5

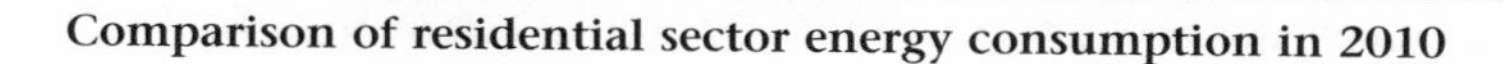

Comparison of residential sector energy consumption in 2010

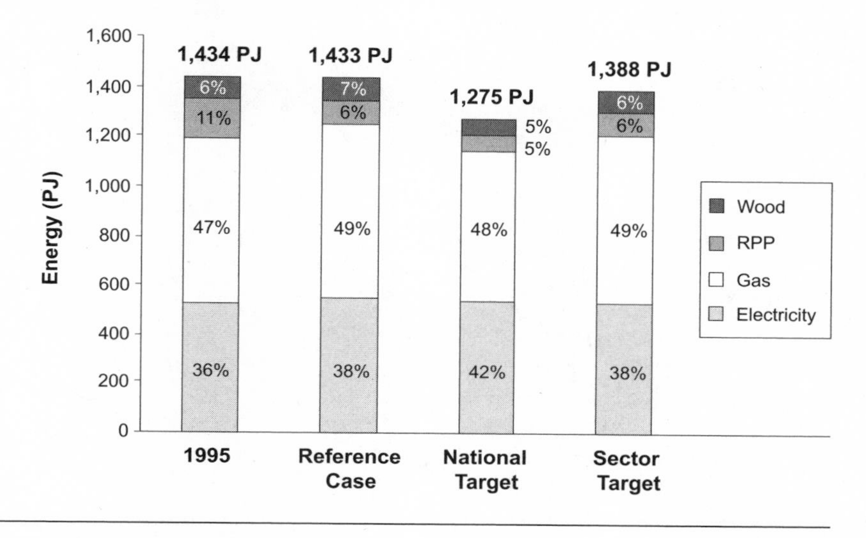

efficiencies as high as 94%, significantly higher than the current average (which in 1996 was 66% for gas-heated equipment stock).

- Reduce hot water demand by installing devices that reduce water flow (e.g., efficient shower heads).

The importance of these reduction sources and fuel switching are shown in Figure 5.6. In addition, emission reductions occur from changes to future urban form ("community energy management" shown in Figure 5.6). More compact patterns of housing development, with a greater share of multistoried buildings, reduce the amount of energy used for heating due to shared walls. However, overall change to urban form is limited by the short simulation period. With a longer time horizon, urban form changes can have a significant impact on emissions.

Reductions in electricity consumption are relatively small (5 PJ in the national target simulation). They are achieved principally through efficiency improvements to the building shell, lighting, water heating, and appliances, as well as through increased penetration of *heat pump* technologies, which transfer heat between the outside air or ground and the building. The importance of each action is shown in Figure 5.7. The community energy management action also contributes to electricity savings (as it does to direct emission reduction savings) for housing that is electrically heated.

Figure 5.6

Sources of residential sector direct GHG emission reductions in 2010 (national target)

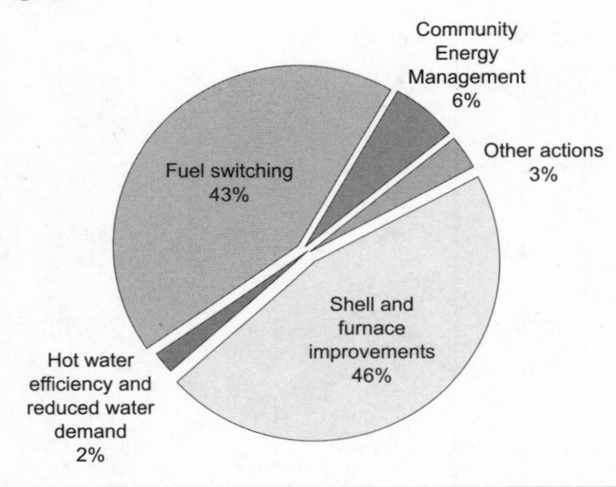

Figure 5.7

Sources of residential sector electricity reductions in 2010 (national target)

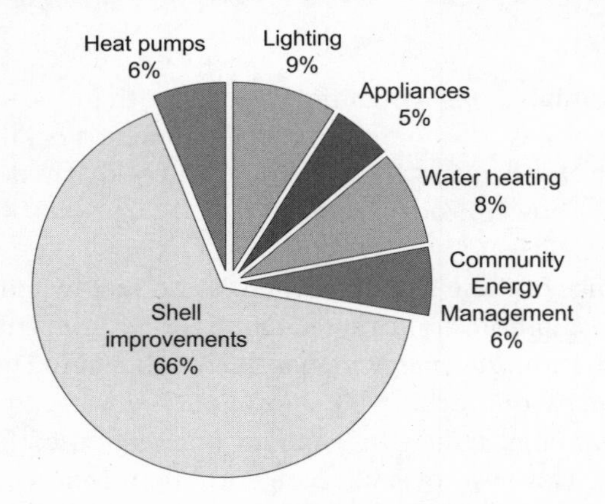

Costs

Table 5.4 shows residential costs for the sector target and the national target. Under the sector target, the residential sector shows a net benefit compared to a net cost under the national target. The change occurs because, under the national target, the sector contributes more to overall reductions up to the point where its marginal cost equals the economy-wide marginal cost of reduction. This brings in actions with significantly higher capital costs, such as solar hot water heating and heat pumps. The higher price of

Table 5.4

GHG abatement costs, residential sector

Type of target	Total cost (billions $1995)	Marginal cost ($/t CO_2e)
Sector	-0.09	0
National	10.63	120

Note: Total costs shown reflect the net present value of costs from 2000 to 2022.

electricity, which is passed down to the demand sectors, also counteracts the savings in energy costs that occur from actions. The residential sector consumes a large share of electricity relative to most other sectors, and actions to reduce emissions have a negligible effect on reducing electricity consumption.

Implications for Households

In Chapter 4 we discussed how meeting the reduction target could affect the energy prices that households face. We elaborate this issue here. As we noted earlier, the changes that consumers see to their utility bills will depend on the type of policy used. The cap and tradable permit approach increases the delivered price of energy, thus influencing households to use energy more efficiently and to switch to fuels that contain less carbon. Figure 5.8 shows predicted electricity prices for the reference case and the national

Figure 5.8

Residential electricity prices in 2010

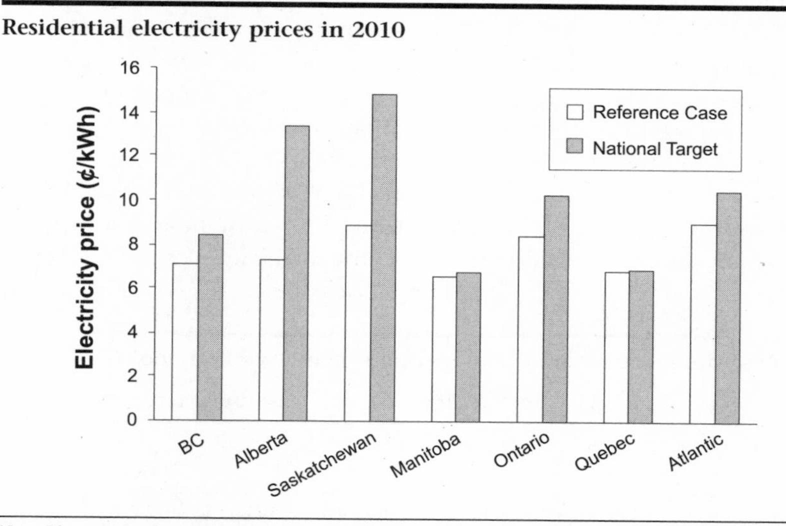

Note: The national target prices include the electricity costs passed on in price as well as the emissions-trading permit price.

Figure 5.9

Residential natural gas prices in 2010

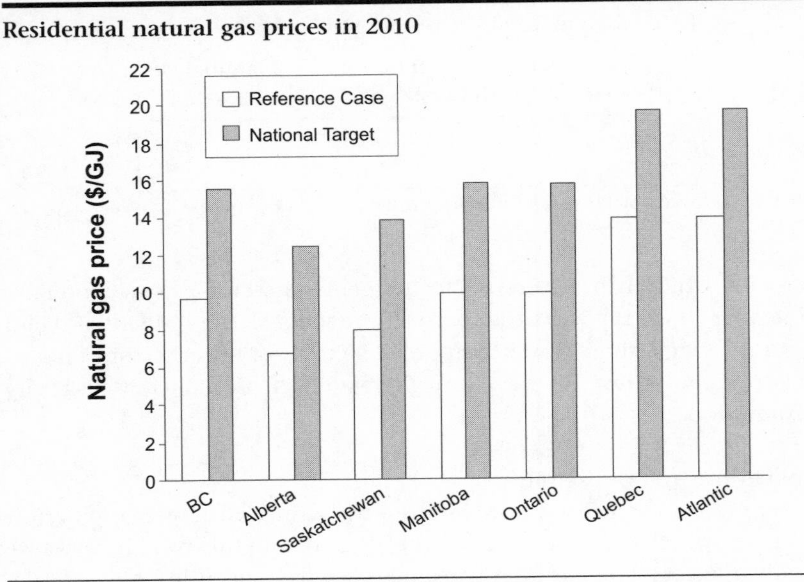

target simulation. Figure 5.9 shows natural gas prices. Impacts vary by region due to differences in fuel mixes – Alberta, Saskatchewan, and the Atlantic region have more carbon-intensive electricity production and thus pass on a greater permit price to the residential consumer. Increases in the price of electricity are greater than those of natural gas in Alberta and Saskatchewan; the reverse is true for other regions.

Because we assume that the tradable permit system is a combination of auctioning and grandfathering, net financial flows in the economy due to permit sales are returned to households as a proportion of permit fees that they are paying. Thus, there are no additional costs to households from the permit system itself.

The total costs shown in Table 5.4 can also be expressed as annual household and per capita costs. These costs are shown in Table 5.5 for both policy simulations. In addition to investment and energy costs, these figures

Table 5.5

Annual residential costs, per person and per household

	Sector target ($)	National target ($)
Per person	-0.42	51.05
Per household	-1.11	135.88

Note: Based on projected population and household figures in 2010 (in *CEOU*). Costs are calculated based on equal annual payments over 22 years.

represent costs associated with trying out new and different technologies, such as higher risks of failure as well as higher information, acquisition, and installation costs. Losses of consumers' value are also included. Under the sector target, the energy efficiency programs save household expenses, since these investments result in energy savings that are greater than capital and other costs.[3]

Commercial/Institutional Sector

The commercial/institutional sector encompasses a wide range of activities and buildings. They include institutions such as hospitals and schools and commercial spaces such as offices, retail stores, and warehouses. Together, retail and office buildings consume almost 50% of total energy in the sector, followed by schools and hospitals. Direct GHG emissions stem from energy use for space heating, water heating, and cooking, while electricity use for cooling, lighting, ventilation, office equipment, and refrigeration is a source of indirect emissions. The energy use of a building is influenced by the amount of floor space, the location, and the type of activity housed. The initial design and construction strongly influence overall energy efficiency over the life of the building. Many components, especially those that are part of the building shell, will not be revised for 25 to 30 years. Gradually, the overall energy intensity of equipment stock will change as new buildings are outfitted with more energy-efficient equipment and as old, worn-out equipment is replaced in existing buildings.

Another important source of GHGs within the institutional sphere is landfill gas, a product of the anaerobic decomposition of organic wastes in landfills. Landfill gas is composed of approximately 50% carbon dioxide and 50% methane. The latter is a potent GHG, with 21 times the global-warming potential of carbon dioxide. Landfill gas can be collected through a series of wells and piping systems installed in the landfill sites and either flared or utilized. Only about a quarter of all landfill sites collected the gas in 1997.[4]

Emission Reductions

Emission reductions in the commercial/institutional sector are described in Figure 5.10 and Table 5.6. Under a sector target, 7.7 Mt CO_2e are reduced. Relatively few reductions are made because reference case emissions are not projected to be significantly higher than 1990 levels in 2010 (growth in commercial floor space, the main determinant in GHG emissions, is relatively slow). When the sector targets are no longer applied, this sector contributes more emission reductions (from 7.7 to 10.5 Mt CO_2e) by drawing on additional cost-effective opportunities, while sectors such as transportation and industry contribute less. Nevertheless, even in this case the sector is a relatively small contributor to total national emissions reductions (6%).

Figure 5.10

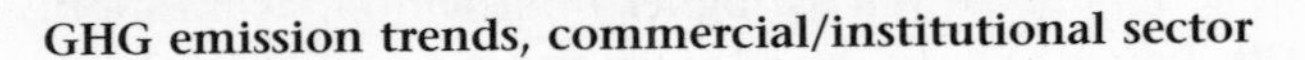

GHG emission trends, commercial/institutional sector

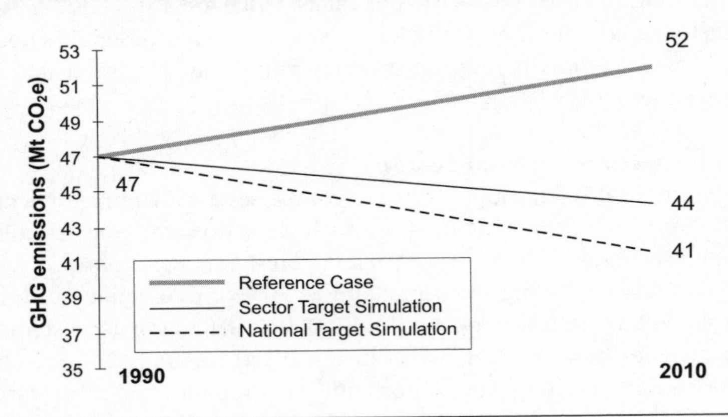

Table 5.6

GHG emission reductions, commercial/institutional sector

	National target	Sector target
Change relative to reference case in 2010 (Mt CO_2e)	-10.5	-7.7
Percentage change relative to reference case in 2010	-20%	-15%
Percentage change in emissions relative to 1990 level	-12%	-6%
Contribution to national target reductions (% of total)	6%	5%

Sources of Reductions

The single most important action in this sector occurs from collection and use of landfill gas. New and expanded well and piping systems at 50 landfill sites achieve a reduction of 6 Mt CO_2e per year, which is between 60% and 80% of total emission reductions from this sector. Using landfill gas to generate electricity has the additional benefit of offsetting GHG emissions from other power sources but is more expensive and is only employed under the national target.

Figure 5.11 relates the capture and use of landfill gas to other actions that contribute to direct emission reductions. Given that space heating contributes significantly to emissions in the sector, building shell and *HVAC system* (heating, ventilation, and air conditioning) improvements are the next largest source of reductions. More specifically, reductions occur from the following.

Figure 5.11

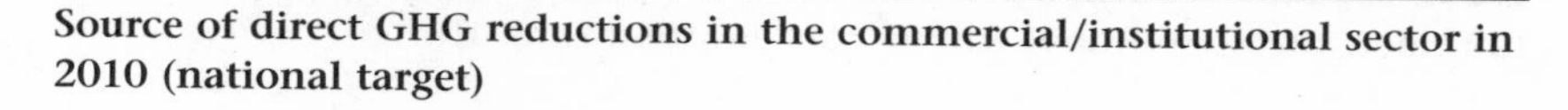

Source of direct GHG reductions in the commercial/institutional sector in 2010 (national target)

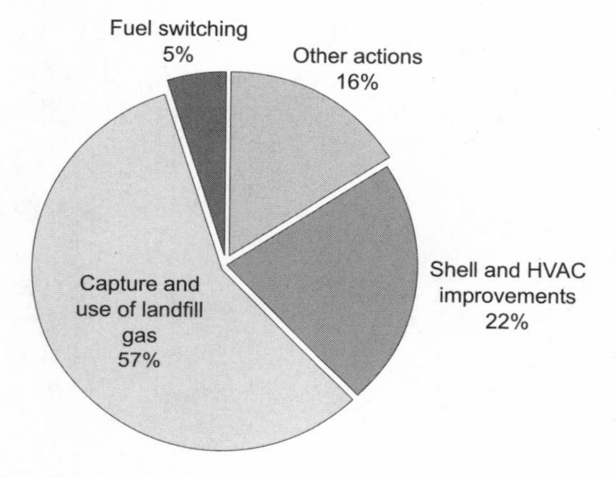

Note: Emission reductions are offset by cogeneration, which increases direct emissions.

- Decreasing the heat demand by using temperature setbacks and improving the building shell (adding insulation to walls and roofs and changing window design and type).
- Improving boiler efficiency – using condensing boilers, which operate at seasonal efficiencies in the 90% range.
- Improving operation and maintenance practices.

Other actions that reduce direct emissions include the use of water boilers with higher efficiencies and the use of water-saving devices. (These actions are indicated by Other Actions in Figure 5.11.)

Fuel switching is responsible for the remainder of the direct emission reductions; commercial buildings switch away from oil and gas toward electricity. Changes in energy use patterns can be seen in Figure 5.12, which shows consumption of different types of energy in the reference case and policy simulations. The figure also indicates the effectiveness of the energy efficiency actions described above. The use of natural gas, oil, and liquefied petroleum gas declines significantly – by approximately 18% from the reference case under the national target (from 590 PJ to 499 PJ) and by 9% under the sector target (from 590 PJ to 536 PJ).

Total electricity decreases by about 12% between the reference case and the national target, from 52 PJ to 44 PJ, and even further under the sector target. As shown in Figure 5.13, this reduction is mainly due to the impact

Figure 5.12

Commercial/institutional energy consumption in 2010

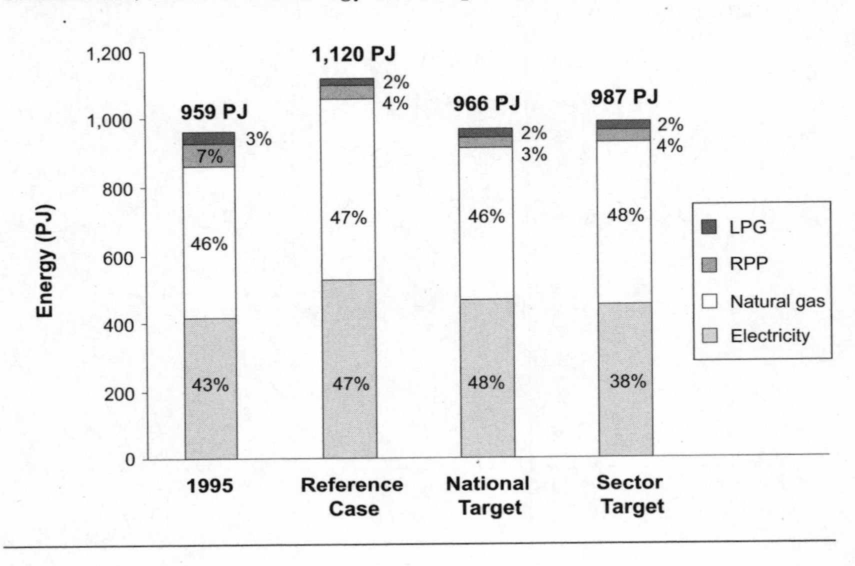

Figure 5.13

Source of electricity reductions in the commercial/institutional sector in 2010 (national target)

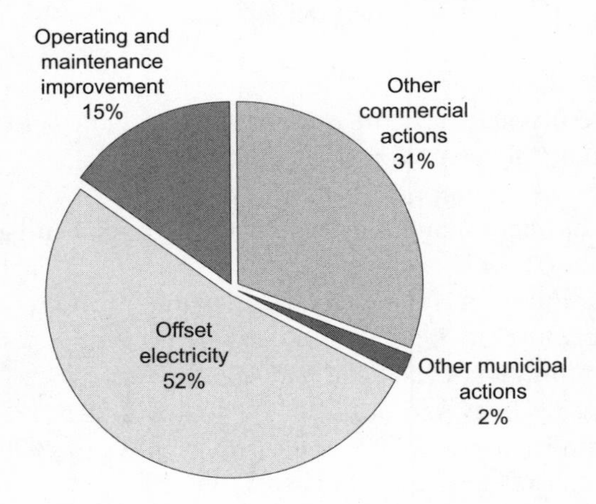

of electricity generated within the sector from tapping landfill gas and through cogeneration in large buildings such as hospitals and universities, as well as that which is generated as part of district energy systems.[5] Under the national target, cogeneration generates 11,420 gWh (9% of electricity used by the sector) in 2010, and landfill gas generates 879 gWh (1% of total electricity used). This is important in regions where this cogenerated electricity displaces fossil fuel-based generation from the electricity sector.

Other reductions in electricity consumption occur from improved operating and maintenance practices, as well as from other commercial and municipal actions. "Other Commercial Actions" include shell improvements and use of temperature setbacks (where room temperature is adjusted according to occupancy), improved air conditioning efficiency, improved lighting design and efficiency, and improved hot water heating efficiency. "Other Municipal Actions" refer to actions by municipalities to divert solid waste and to conserve water.

Costs

The commercial/institutional sector shows a net benefit under the sector target and a net cost under the national target (Table 5.7). More reductions occur in the second case because the sector faces a marginal cost of reduction of \$120/t CO_2e (the economy-wide permit-trading price) rather than \$10/t CO_2e under the sector target.

Commercial/institutional actions generate substantial energy savings. In particular, the generation of electricity from landfill gas and from cogeneration offsets expensive purchased electricity (whose prices reflect actions to the electricity sector). The sector's contribution to total national costs matches its contribution to national emission reductions; both are about 6%.

As in the residential sector, changes in consumers' value under the policy simulation are important information for a policy maker. In the commercial/institutional sector, they may represent preferences for the quality of incandescent lighting over fluorescent lighting and/or for the ventilation quality of a particular HVAC system.

Table 5.7

GHG abatement costs, commercial/institutional sector

Type of target	Total cost (billions \$1995)	Marginal cost (\$/t CO_2e)
Sector	-2.76	10
National	2.71	120

Note: Total costs shown reflect the net present value of costs from 2000 to 2022.

Transportation Sector

Transportation is integral to the way in which our modern economy functions, moving goods across the country and outside our borders and allowing people to be mobile. Substantial energy is required to do these tasks – almost the same amount of energy required to manufacture all goods in the Canadian economy.[6] The transportation sector currently generates about one-quarter of Canadian GHG emissions. Personal vehicles contribute slightly more than half of these emissions, of which automobile emissions comprise 42% and light-duty trucks 17%. The number of passenger vehicles is growing, as is the distance that each vehicle is being driven. Of the remaining sources, heavy-duty trucks and buses contribute 20%, aircraft 9%, rail 4%, marine 3%, and other activities 6%.[7]

In the reference case, emissions increase 35%, which along with electricity is the largest among all sectors. The current outlook for car and light-truck emissions, based on projections of passenger car sales, new car fuel efficiency, and the overall vehicle stock fuel efficiency suggests that anticipated fuel efficiency gains will continue to be outweighed by traffic growth: increases in vehicle stock and the number of kilometres driven. Furthermore, a greater number of people are purchasing sport utility vehicles relative to other vehicle types. Between 1997 and 2010, an increase in 14.7 Mt CO_2e emissions is expected from Canadian passenger vehicles. Of this total, an increase of only 0.75 Mt CO_2e is expected for passenger cars, while gasoline trucks, under which sport utility vehicles are classified, see an increase in emissions of 14 Mt CO_2e. Gasoline truck efficiency is not expected to improve as much as car efficiencies over this time period. Trends in freight also contribute to growth in the reference case, with the greatest growth in air traffic and trucking.

Emission Reductions

Figure 5.14 and Table 5.8 show how this sector contributes to emission reductions under a sector target and the national target. Although emissions in the reference case grow substantially, the transportation sector is capable of considerable emission reductions and is a huge contributor to total reductions in the modelling results (between 21% and 38%, depending on the simulation). Under a sector target, 58 Mt CO_2e are reduced, decreasing to 44 Mt CO_2e under the national target. The transportation sector contributes less under a national target because other sectors, such as commercial/institutional and residential, have opportunities that are cost-effective in comparison to the transportation measures that cost more than \$120/t CO_2e.

Sources of Reductions

Emission reductions in the transportation sector occur through a decrease in energy consumption, which results from changes in transportation modes,

Figure 5.14

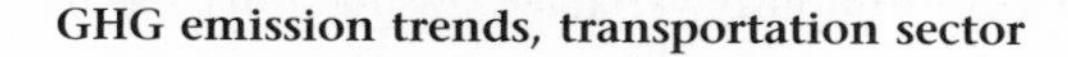

GHG emission trends, transportation sector

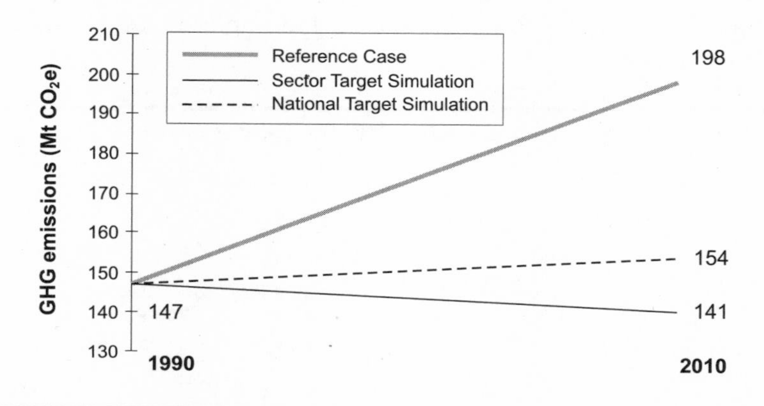

Table 5.8

GHG emission reductions, transportation sector

	National target	Sector target
Change relative to reference case in 2010 (Mt CO_2e)	-43.7	-57.6
Percentage change relative to reference case in 2010	-22%	-29%
Percentage change in emissions relative to 1990 level	5%	-6%
Contribution to national target reductions (% of total)	24%	38%

improved vehicle efficiencies, and reductions in travel demand. Changes in total energy use are shown in Figure 5.15. Fuel switching is negligible, and almost all reductions occur in the use of refined petroleum products (principally gasoline and diesel). Other types of energy (ethanol and electricity), which are a small share of energy consumption, do not change significantly, though both increase slightly in the policy simulations due to greater use of electric public transit in metropolitan areas and greater ethanol use in passenger vehicles.

Emission reduction actions are shown in Figure 5.16 (for the national target). Key actions are fuel-efficiency improvements to new vehicles, public transit enhancements, and improved fuel efficiency from reductions in travelling speeds.

- *New vehicle efficiency:* The average fuel consumption per kilometre travelled for cars and light trucks declines by 25% by 2010. Improving car efficiency can be achieved by reducing engine friction, lowering aerodynamic drag, and employing continuously variable transmissions, lightweight interiors, high-strength steel bodies, and higher engine compression ratios (among others). This action alone reduces emissions by 5.2 Mt CO_2e.

Figure 5.15

Petroleum product consumption in 2010, transportation sector

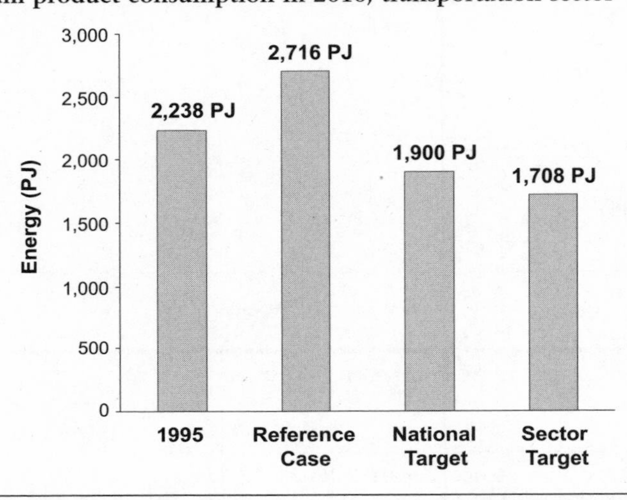

Note: The quantity of nonpetroleum products (compressed natural gas and electricity) is negligible: 14 PJ in the reference case, 14 PJ under the national target, and 21 PJ under the sector target.

Figure 5.16

Source of transportation emission reductions in 2010 (national target)

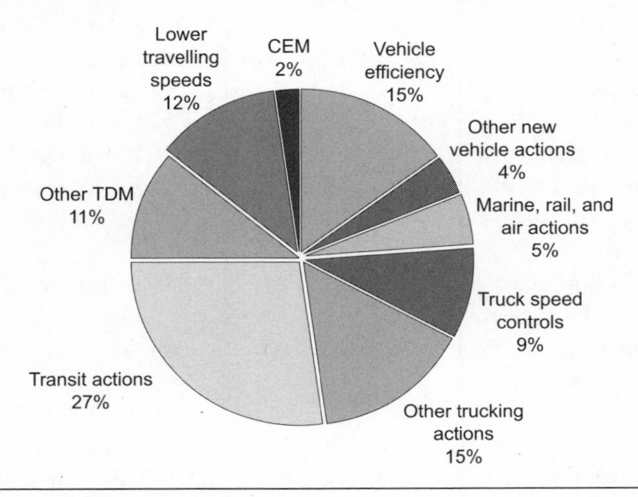

- *Public transit enhancements:* A 9.5 Mt CO_2e reduction occurs from shifting auto use to transit by increasing the availability of fast, convenient, and reliable transit service and creating a greater incentive to use transit through pricing structures. Transit is enhanced by both increased infrastructure (commuter rail, light rail, heavy rail, and grade-separated bus lanes) and improved service (more frequency, new routes, improvements to safety, and convenience). Transit shares across Canada increase by approximately 2%.
- *Reduced highway speeds:* Travelling at the posted national highway speed limits, rather than exceeding them, reduces emissions by 4.2 Mt CO_2e by increasing fuel efficiency.
- *Truck fuel speed monitoring:* The wider use of vehicle monitoring and control systems on engines (with road speed and idling limits) allows for more fuel-efficient driving and thus a reduction of 3.2 Mt CO_2e.

Other activities that reduce emissions include the following.

- *Trucking actions:* This strategy includes retiring older, inefficient trucks and improving lubricants.
- *Other new vehicle and fuel-switching actions:* Ethanol fuels are made more available, and alternative fuel technologies – such as natural gas, electric hybrid, and hydrogen fuel cell – are incorporated into vehicles.
- *Marine, rail, and air actions:* A number of short-term air actions are effective, such as more fuel-efficient airport approaches and changing aircraft routing.
- *Other actions directed at passenger transportation:* Examples include introducing more high-occupancy vehicle lanes, telecommuting, car sharing, improving the fuel economy of the on-road vehicles (e.g., through synchronized traffic signals), and using intelligent transportation systems technologies.

Changes in urban land use from community energy management also contribute to transportation emission reductions (as they do to residential emission reductions), but the tight Kyoto time frame limited the impact of this action. Urban form can have significant impacts on transportation in the longer term by influencing the relative use of public transit versus automobiles, encouraging walking and cycling, and reducing travel distances.

Costs

The transportation sector bears substantial costs for meeting the reduction target, although the degree differs according to whether reductions are made under a sector or a national target. Table 5.9 shows costs to the transportation sector, which are a third lower under the national target, reflecting

Table 5.9

GHG emission abatement costs, transportation sector

Type of target	Total cost (billions $1995)	Marginal cost ($/t CO_2e)
Sector	28.96	$300/t
National	19.47	$120/t

Note: Total costs shown reflect the net present value of costs from 2000 to 2022.

lower emission reductions. The high-cost measures for this sector are not implemented, and greater reductions occur in other sectors at lower overall cost to the economy. Even though the transportation sector contributes a quarter of reductions, it incurs nearly one-half of all costs.

Implications for Lifestyle and Competitiveness of the Sector

Gasoline Prices and Annual Costs

In the national target simulation, the permit-trading price increases the delivered price of gasoline (and other fuels). Figure 5.17 shows gasoline prices for different regions for the reference case and the national target simulation.

Figure 5.17

Gasoline prices in 2010

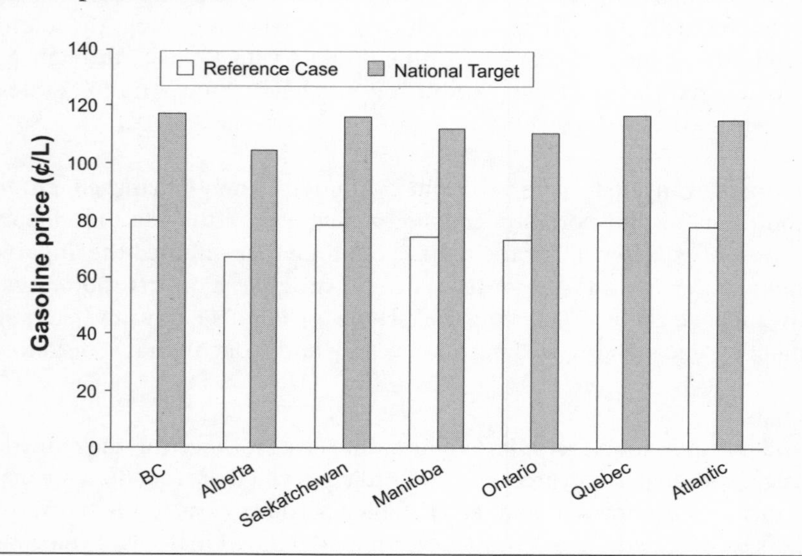

Table 5.10

Increase in annual transportation costs, per person and per household

	Sector target ($)	National target ($)
Per person	139.05	93.49
Per household	370.06	248.82

Note: Based on projected population and household figures in 2010 (in *CEOU*). Costs are calculated based on equal annual payments over 22 years.

The cost implications of the simulated policies can also be expressed as annual costs. These costs are shown in Table 5.10 and assume equal payments over 22 years. The per person cost for actions under the sector target (delivered through various programs) are higher because, in this case, the transportation sector contributes more to emission reductions.

Consumers' Value

Estimates of consumers' value are quite important to cost estimates for the transportation sector. The costs reported in Table 5.9 increase by about $8 billion if a fuller estimate of consumers' value losses (beyond the Issue Table's estimate) is included. (See the sensitivity analysis of the last chapter.)

Impact on Transportation-Related Businesses

According to the macroeconomic results, there will be positive business impacts due to increased investment in urban transit infrastructure and operating vehicles but a moderate reduction in personal vehicle production associated with increased transit use. This latter effect is highly uncertain, though, because it is unknown whether higher operating costs for personal vehicles will lead to reduced vehicle acquisition and thus reduced production. Some analysts argue that, even if people commute more by public transit, and thus drive less, they will still own roughly the same number of vehicles, the personal vehicle being desired for its many other attributes. Others argue that greater public transit services will allow for more vehicle sharing both within and between households, a trend that is emerging in Europe. This trend has very different implications for production of vehicles and thus the macroeconomic effect.

Electricity Sector

Most electricity in Canada is generated from sources that have negligible GHG emissions in operation: hydro generates 62%, nuclear 11%, and other renewable sources (wind, biomass, solar) 1%. Nevertheless, the remaining

Figure 5.18

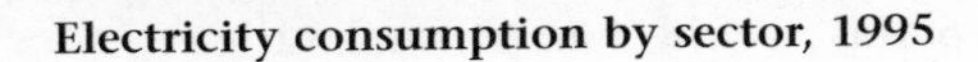

Electricity consumption by sector, 1995

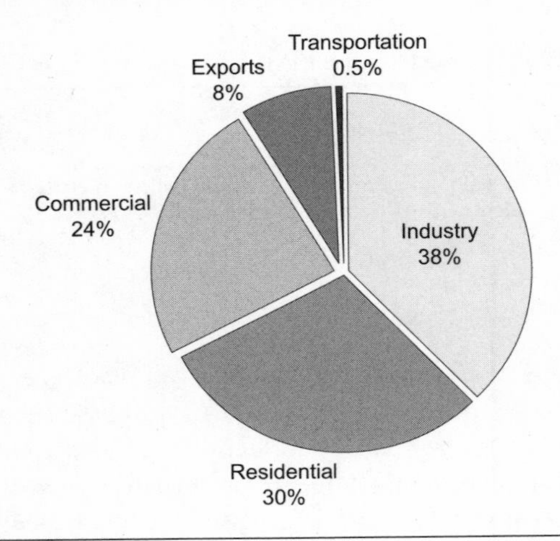

generation from the combustion of fossil fuels – coal (19%), natural gas (5%), and oil (3%) – produces 124 Mt CO_2e, 18% of Canada's total emissions.[8] This is more than all direct emissions in the residential and commercial/institutional sectors. Electricity production is based on electricity use by demand sectors in Canada, as well as exports to the United States. Figure 5.18 shows electricity consumption shares.

There are significant regional variations in generation. For example, conventional thermal generation (mainly coal based) accounts for the majority of installed capacity in Alberta, Saskatchewan, and New Brunswick, while Newfoundland, British Columbia, Quebec, and Manitoba are primarily hydro based. Electricity exchange between provinces is limited; however, Manitoba, British Columbia, and Quebec have significant exports to the United States (about 8% of electricity generated). Much of Canada's generating capacity came into service after 1970 and has significant life remaining, thus constraining opportunities for emission reductions. The reference case predicts that existing coal generation will be replaced by gas generation and by refurbished coal plants. Nuclear generation will decline overall, while hydro will increase.

Emission Reductions

Emissions in the reference case and policy simulations are shown in Figure 5.19 and Table 5.11. Reference case emissions in the electricity sector increase by 33 Mt CO_2e between 1990 and 2010, a 34% increase. Many relatively low-cost opportunities are available to easily meet the sector's target

Figure 5.19

GHG emission trends, electricity sector

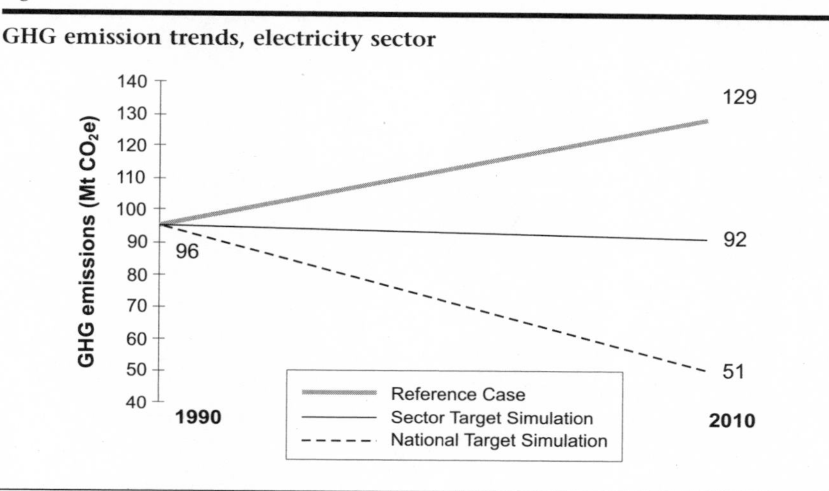

Table 5.11

GHG emission reductions, electricity sector

	National target	Sector target
Change relative to reference case in 2010 (Mt CO_2e)	-77.5	-36.8
Percentage change relative to reference case in 2010	-60%	-29%
Percentage change in emissions relative to 1990 level	-46%	-6%
Contribution to national target reductions (% of total)	43%	24%

reduction of 36 Mt CO_2e, approximately a third lower than reference case levels, at a marginal cost of \$30/t CO_2e. Under a national target, reductions in this sector displace more expensive actions in other sectors (industry and transportation), so that in 2010 emissions are 60% lower than in the reference case. These substantial reductions contribute the largest share, 43%, to total Canadian emission reductions under a national target.

Sources of Reductions

Emission reductions in this sector occur from changes in demand for electricity, fuel switching, and CO_2 capture and sequestration. The relative importance of each is indicated by Figure 5.20 for the national target simulation. Changes in electricity demand, which varies little from the reference case, are not a significant contributor. Under the sector target, electricity demand rises, while under the national target it decreases. The different response occurs because under the sector target the electricity supply sector faces a

Figure 5.20

Source of electricity supply GHG emission reductions in 2010 (national target)

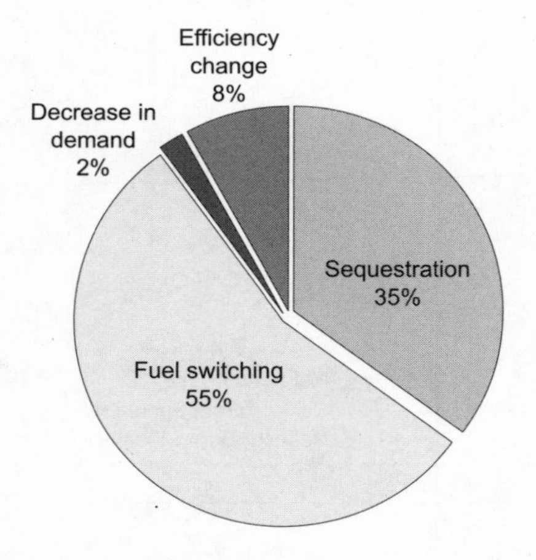

lower marginal cost of emission reductions ($30/t CO_2e) than most demand sectors (see Table 5.1). Thus, for the demand sectors, the cost of the electricity sector's emission reductions passed on in the electricity price is relatively less expensive than making direct emission reductions themselves, and fuel switching to electricity occurs. Overall demand for electricity increases. Under the national target, demand and supply sectors are faced with the same marginal cost for emission reductions, and overall fuel switching to electricity does not occur. High electricity prices instead lead to a drop in demand. However, this drop accounts for only a small decrease in GHG emissions in the electricity sector because switching toward electricity occurs in those regions where it is primarily generated from non-GHG-emitting sources.

Instead, the majority of emission reductions in the electricity sector occurs from fuel switching in generation: oil and coal consumption drops, while generation from natural gas, wind, and hydro increases (see Figure 4.5). Generation from wind is still relatively minor, less than 1% of total generation, though it is double relative to the reference case. The capture and sequestration of CO_2 from coal combustion accounts for another 37% of the reduction. With this action the substitution away from coal use is slowed.

Under the sector target, CO_2 capture and sequestration accounts only nominally for emission reductions because the target can be met at a lower cost through fuel switching without resorting to this action, which is more expensive. Conventional coal generation with CO_2 capture and sequestration costs approximately \$45/GJ, while combined cycle natural gas generation is about \$14/GJ.[9] In the national target simulation, 27.3 Mt CO_2e are captured and sequestered, versus 0.8 Mt CO_2e in the sector target simulation. Instead, fuel switching, particularly away from coal, accounts for almost all of the reductions. The lower marginal cost of the sector target also limits the adoption of wind and biomass generation, since they are higher-cost options relative to natural gas generation.

Since the single action of CO_2 capture and sequestration figures strongly in the results, we discuss it in more detail. In Chapter 2 we described a number of variations in this technology. This analysis incorporates only one option considered to be feasible within Canada at this time: CO_2 is removed from flue gases in coal-fired electricity production and stored underground in deep saline aquifers in Alberta and Saskatchewan at depths of more than 800 metres, a method that enables the storage of CO_2 in a dense, supercritical form. Capture occurs through the use of molecular sieves, physical and/or chemical absorption, and membranes. This option for CO_2 storage has already been applied commercially off the coast of Norway. The worldwide potential for CO_2 storage in such aquifers is thought to be thousands of gigatonnes – enough for the sequestration of several hundred years of CO_2 from fossil fuel combustion.

Costs

Table 5.12 shows costs in electricity supply for the sector and the national targets. Under the sector target, costs mainly reflect the extra outlay of money to switch to small hydroelectric and natural gas generation from coal-fired plants, which are less costly to build and operate. Costs are higher under the national target because emission reductions are much more significant. These costs include the use of more renewables and GHG capture and sequestration.

When the electricity sector reduces the amount of electricity that it generates because of reduced demand (as under the national target), the sector saves money in investment, operating, maintenance, and energy costs. Reduced generation also leads to a reduction in revenue within that sector, an expense that needs to be accounted for in total costs. The reverse is true when demand increases. These revenue changes are captured as a demand adjustment in Table 5.12.

All of the costs shown are passed on to the demand sectors in changes to electricity price. Thus, the cost tables for all other sectors in this chapter incorporate the costs shown here.

Industrial Sector

The industrial sector, which includes mining, construction, and manufacturing activities, is the largest GHG-producing sector in Canada. The sector

Table 5.12

GHG emission abatement costs, electricity sector

Type of target	Total cost (billions $1995)	Marginal cost ($/t CO_2e)
Sector	15.71	30
National	28.47	120

Note: Total costs shown reflect the net present value of costs from 2000 to 2022.

produced 340 Mt CO_2e of direct GHG emissions in 1998, the majority of which are energy consumption based.[10] Energy is particularly critical in the production of basic industrial products, such as cement and steel, which are used to produce goods for final consumption, either within or outside Canada. These primary products industries, often referred to as Tier 1 industries, account for more than 80% of total industrial energy consumption. They include iron and steel, pulp and paper, metal smelting, petroleum refining, chemical manufacturing, and industrial minerals (cement and lime production). Mining accounts for another 5% of energy consumption. The remaining industries, which are numerous and diverse (food processing, transportation equipment manufacturing, etc.), use relatively little energy, 16%, but are responsible for 60% of industrial economic output as measured by GDP.[11]

In the reference case, industrial emissions grow considerably over 1990 levels, 28% from 1990 to 2010 (from 199 Mt CO_2e to 255 Mt CO_2e). This growth occurs because the production in a number of energy-intensive subsectors is expected to grow significantly. Figure 5.21 gives a sense of which industrial subsectors contribute most to emissions as projected into the future.

The "Fossil Fuel Industries" generate by far the largest quantity of GHG emissions. This grouping in the figure refers to the *upstream* component of the oil and gas industry (exclusive of petroleum refining, which is considered *downstream*) and includes emissions from oil and natural gas exploration, production, flaring, drilling, well service, gas distribution, upgrading (including oil sands bitumen), and pipelines. Emission increases in this subsector are related to strong growth in oil and gas exports to the United States. Many upstream emissions are *fugitive emissions* (51 Mt CO_2e in 1998), the intentional or unintentional releases of GHGs from the production, processing, transmission, storage, and delivery of fossil fuels. Releases include some carbon dioxide, but the bulk is methane. Emissions from processing bitumen from oil sands to make synthetic crude oil are another large component of this category; oil sands operators released 13 Mt CO_2e in 2000.[12]

Figure 5.21

Reference case emissions by industrial subsector

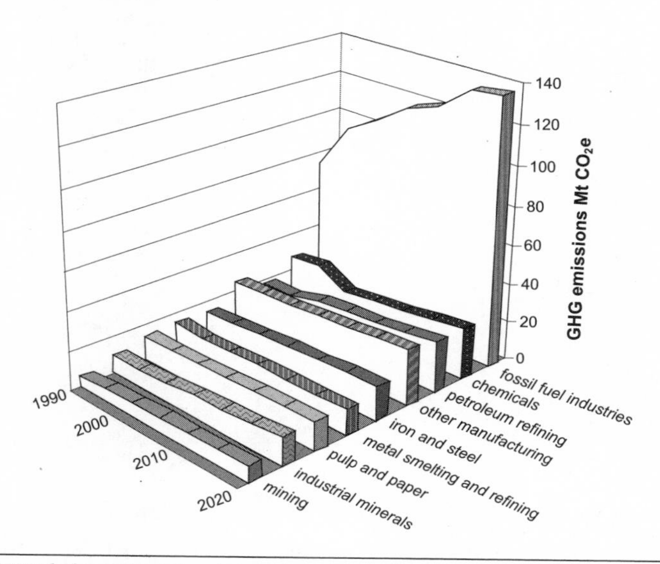

Note: Mining includes coal mining; *pulp and paper* includes forestry; *fossil fuel industries* does not include petroleum refining.
Source: National Climate Change Process, Analysis and Modelling Group, *Canada's Emissions Outlook: An Update* (Ottawa: National Climate Change Process, 1999).

Many "Other Manufacturing" industries require heat (some of which is simply space heating), primarily in the form of steam. As these industries grow, their level of emissions grows as well. This collage of industries spans divergent subsectors from electronics to car manufacturing, food processing, rubber production, textiles and clothing, furniture manufacturing, printing, and others.

Figure 5.21 indicates that emissions decline significantly in chemical product manufacturing. This is primarily due to a single company's action to reduce its release of N_2O emissions.[13] The ability to reduce electricity consumption (and reduce indirect emissions) is important to this subsector. For instance, significant amounts of electricity are used in the electrolytic production of inorganic chemicals. However, since Figure 5.21 represents direct emission only, these emissions are not recorded.

Petroleum refining refers to establishments that manufacture a family of petroleum products. These facilities are referred to as the *downstream* component of the fossil fuel industry. Refineries use up to 7% of their crude input by volume as fuel to separate the crude into fractions and to refine the various petroleum products. Emissions are restricted to production and do not include those generated when the products are actually used.

Table 5.13

Industrial subsector models in CIMS

	Region						
Sector	BC	AB	SK	MB	ON	PQ	AT
Chemicals	√	√			√	√	
Coal mining	√	√	√				√
Industrial minerals	√	√			√	√	√
Iron and steel					√	√	
Mining	√		√	√	√	√	√
Metal smelting and refining	√			√	√	√	√
Natural gas extraction and transmission	√	√	√	√	√	√	√
Other manufacturing	√	√	√	√	√	√	√
Petroleum extraction and refining	√	√	√		√	√	√
Pulp and paper	√	√			√	√	√

A significant proportion of the GHG emissions generated in the industrial minerals branch of industry comes from the chemical reaction used to produce cement and lime. Both products, the largest energy consumers in the industrial minerals, release between 0.5 and 0.7 tonnes of CO_2 per tonne of product.

In Figure 5.21, emissions from most industries remain relatively flat over time, in spite of production growth. This follows a trend that these industries have established over the past two decades. Some industries have the advantage of moving to biomass fuels (pulp and paper, wood products), while others can move from more to less carbon-intensive fuels (industrial minerals).

Because industry emissions are the result of many distinct processes, CIMS represents these emissions through detailed industrial submodels. Regional submodels exist for industries that consume significant amounts of energy; they are shown in Table 5.13. The "Other Manufacturing" submodel includes all industries that are not energy intensive enough to be described by distinct models (e.g., food manufacturing, textiles), as well as regional industries that are too small to be modelled separately (e.g., only small quantities of steel are produced in most provinces).

Emission Reductions

Figure 5.22 and Table 5.14 show how the industry sector contributes to GHG emission reductions in order to meet a sector target and a national target. The figure shows that none of the policy simulations brings emission levels below those of 1990. As indicated by the "Sector Target Simulation"

Figure 5.22

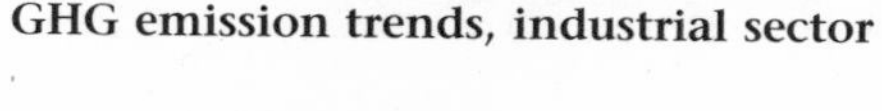

GHG emission trends, industrial sector

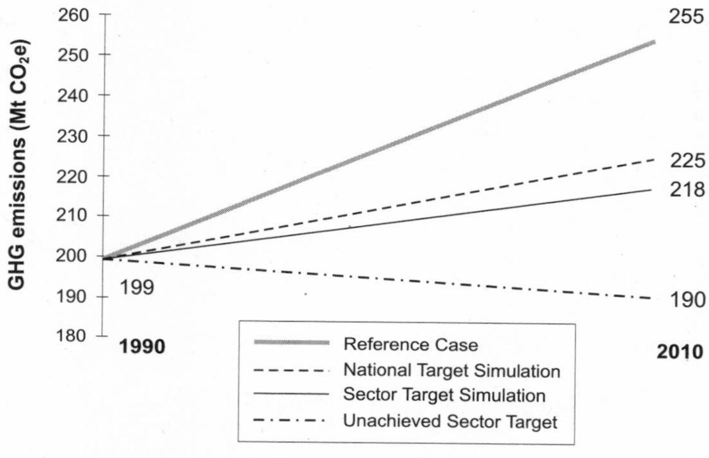

Table 5.14

GHG emission reductions, industry sector

	National target	Sector target
Change relative to reference case in 2010 (Mt CO_2e)	-29.8	-36.8
Percentage change relative to reference case in 2010	-12%	-14%
Percentage change in emissions relative to 1990 level	13%	10%
Contribution to national target reductions (% of total)	16%	24%

line, even including actions with relatively high costs (up to \$300/t CO_2e) can only reduce emission levels to 218 Mt CO_2e, which is still 10% above 1990 levels. The "Unachieved Sector Target" line shows what emission levels would be to reach 6% below 1990 levels. Under a national target, reduction opportunities in other sectors (electricity, residential, and commercial/institutional) displace industrial actions, so that in 2010 emissions are reduced by only 12% from reference case levels. While the percentage share is smaller relative to those of most other sectors, this sector still contributes a sizeable share to total Canadian reductions – between 16% and 24% – due to the overall size of industry emissions.

There are reasons why the industrial sector has a low-percentage reduction from the reference case. First, firms are more likely than households and even commercial tenants to have already pursued cost-effective options

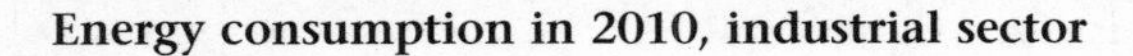

Figure 5.23

Energy consumption in 2010, industrial sector

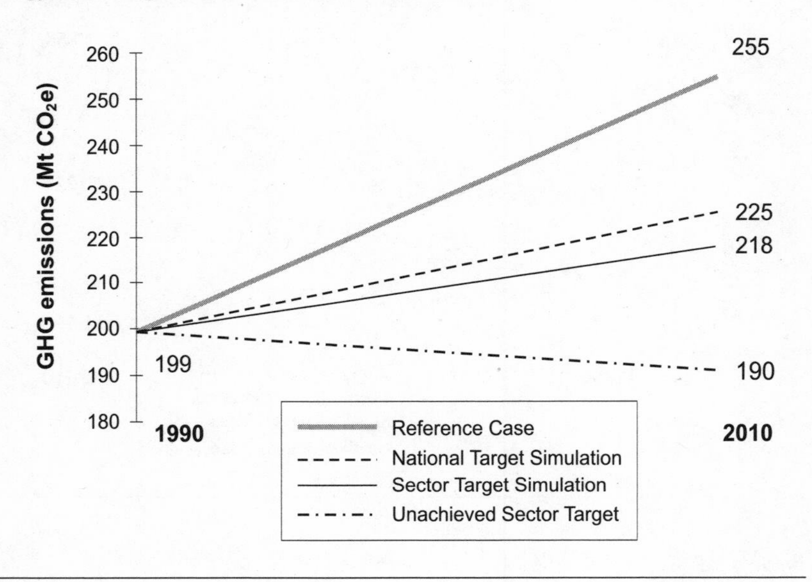

to reduce energy consumption. Thus, there is less additional potential to acquire equipment that cost-effectively improves energy efficiency or switches fuels. Second, some sectors are physically limited in their ability to reduce energy use, in particular fossil fuel use. This inability may be because of minimum thermal requirements of industrial processes (ore refining, pulp production, etc.) or because of minimum energy requirements of generic mechanical energy uses such as conveyance, fans, and blowers.

Sources of Emission Reductions

Emission reductions in this sector are based mainly on improving energy efficiency (32%), fuel switching (20%), and reducing fugitive and nonenergy emissions (48%). Figure 5.23 shows fuel switching and the effect of changes in energy efficiency. Fuel shares stay more or less consistent between the reference case and the policy simulations, although total energy consumed drops by 6% under the national target and 7% under the sector target.

Close to a third of industrial emissions occur from *generic actions,* which are not specific to a particular industry. The remainder occurs from actions that are specific to the processes of a particular industry (*process actions*). Generic actions focus on auxiliary systems that supply energy services to major processing equipment during operation. These auxiliary systems fall into four general categories: steam generation systems (boilers and cogenerators), lighting, HVAC systems, and electric motor systems (pumps, fans, compressors, and conveyors).

In some cases the energy service meets the direct need for steam, pumping, or compression, while in other cases it provides suitable conditions for production to continue – lighting and HVAC systems. The latter is not significant to GHG emission abatement efforts, especially in Tier 1 industries. However, significant reductions can occur through energy efficiency improvements to steam generation systems and to electric motors and their attached auxiliary devices.

The efficiency of steam generation varies greatly depending on boiler design, age, and fuel used. For modern oil and gas boilers, thermal efficiencies may be 85% or higher. Efficiencies can be improved by introducing heat recovery systems and by installing regenerative burners with computerized fuel-air mixtures to maximize fuel efficiency. Substantial energy efficiency improvements can also occur by using cogenerators rather than simple steam boilers. Although this technology reduces overall emissions (both direct and indirect), switching to cogeneration will typically increase an industry's direct emissions due to greater fossil fuel use to generate electricity.

The vast majority of electricity consumed by industry is used by motor systems. Although substantial potential exists to improve the efficiency of electric motors, there is greater potential to improve the efficiencies of equipment driven by them – pumping, air displacement, compression, conveyance, and other types of machine drive. The latter category comprises all electrically driven equipment unique to a given production process.

Historically, pump efficiency has not been a major concern. The technology is mature: the best new pumps available are only 3% to 10% better than the average conventional pump.[14] Replacing valve control with a variable speed drive can improve system efficiency by 20% to 30%; however, most pump systems have already been converted since variable speed drives are used to control processes and for easy maintenance.

Air displacement systems, such as fans and blowers, consume a significant amount of electricity in the industrial sector, typically accounting for 20% of electricity demand. Fan systems often consist of a speed-control device, a motor, a fan, a control vane or damper, and a duct system. There are usually opportunities for efficiency improvements in each of these components and by optimizing the whole system. Although fan technologies are mature – no major design changes have occurred in the past 20 years – there is still room for engineered efficiency improvements. Improved impeller designs and better construction materials may achieve a 10% efficiency improvement over the next 20 years.

A conveyance system is a horizontal or inclined device for moving bulk material. Conveyance systems account for a small portion of industrial electricity demand, typically less than 5%. The simple nature of conveyance

Figure 5.24

Emission reductions by industry subsector (Mt CO_2e, national target)

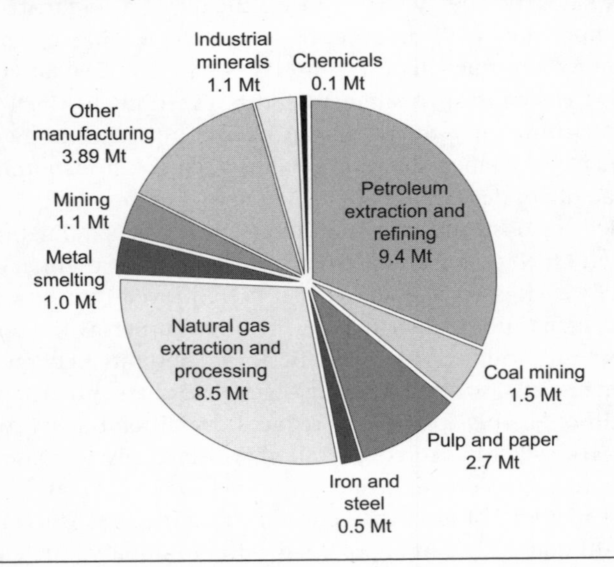

systems means that the potential for increased efficiency is small compared with that of other systems.

Compressor systems are designed to increase the pressure of a gas to a useful level. They are the least efficient auxiliary system: total system efficiency averages between 15% and 20%. This is due to the compressible nature of a gas, which absorbs energy as it is compressed, and the loss of pressure from air leakage. Substantial opportunities for power savings exist.

Emission Reductions by Industry Subsector

Figure 5.24, which shows the contribution of each subsector to total industry sector reductions under the national target, indicates that emission reductions are concentrated in the natural gas extraction and petroleum refining subsectors, which together contribute 60% to total reductions. Emission reductions occur from the following.

- *Natural gas extraction and transmission* (reduction of 8.5 Mt CO_2e, 28% of total industry): Natural gas is extracted, processed, and transported via pipeline to the consumer. At each stage in the process, emissions are released, and abatement actions can be taken. Natural gas transportation offers the most cost-effective reductions from leak detection and repair programs *and by switching to electric-driven compressors*. Other reductions from processing natural gas are more costly. They include replacing over-

Figure 5.25

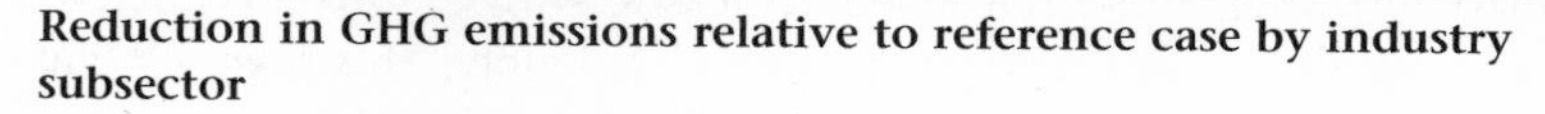

Reduction in GHG emissions relative to reference case by industry subsector

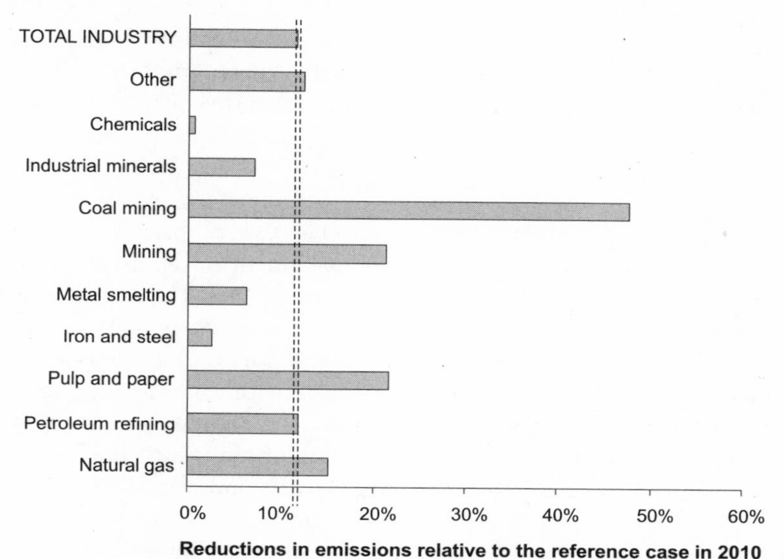

sized compressors for improved efficiency, recovering flare gas using electric drive compression, and altering the sweetening process for sour gas.

- *Petroleum extraction and refining* (reduction of 9.4 Mt CO_2e, 32% of total industry): As in the natural gas subsector, many emission reductions come from reducing upstream fugitive emissions, which account for 60% of emissions in this subsector. Actions can be taken to flare or incinerate gases that would normally escape and to change the way in which flaring is done, such as upgrading to high-efficiency flares or using flare gas to generate electricity. Downstream, actions that save energy in heat and steam provision offer the greatest potential for emission reductions, since significant quantities of heat are required to refine crude oil. Savings occur both from reducing demand for heat and steam services and from more efficient generation and use of heat and steam.

Given the substantial size of emissions produced by the industrial sector, emission reductions of relatively less important subsectors still contribute substantially to overall reductions and should not be overlooked. Figure 5.25, which depicts each subsector's emission reductions as a percentage of reference case emissions, shows that some sectors have significantly more potential to make reductions. The coal-mining, metal- and mineral-mining, and pulp and paper subsectors have significantly higher emission reductions relative to their respective reference cases, while subsectors such as chemicals and iron and steel affect emissions negligibly. This suggests that there may be more cost-effective reductions in the aforementioned industries, while the latter industries can reduce emissions only at great cost. Nevertheless, the scale of emissions produced by different industry subsectors (Figure 5.21) limits many of these industries in their overall contribution to emission reductions. More about the reduction opportunities in each subsector is described below.

- *Chemical products* (reduction of 0.1 Mt CO_2e, 0.3% of total industry): The chemical products industry is heavily dependent on electricity and on natural gas (both as a feedstock and for a fuel). Thus, the opportunity to move away from carbon-intensive fuels is small. The potential of other actions to reduce emissions is also relatively small. Cogeneration offers one opportunity, but the savings are indirect (direct emissions actually increase) and differ significantly by region (it is highest in regions where electricity is more expensive). The reduction of nitrogen oxide emissions from adipic acid production substantially cuts emissions in this sector, but this reduction is already included in the reference case.
- *Coal mining* (reduction of 1.5 Mt CO_2e, 5% of total industry): Because the production of coal in Alberta, Saskatchewan, and the Atlantic region changes as the quantity of coal demanded by the electricity sector changes, a drop in production accounts for a significant share of emission reductions. Considerable reductions also occur from using coal-bed methane to generate electricity. This option exists where methane density in underground coal beds is sufficient to permit extraction and capture to make electricity. Otherwise, there are few opportunities to reduce emissions, particularly if the coal is used directly in electricity generation. Actions include using more efficient auxiliary systems and computerized process optimization. Export-destined coal in British Columbia and in some operations in Alberta can also improve energy efficiency in cleaning, washing, and drying steps.
- *Industrial minerals* (reduction of 1.0 Mt CO_2e, 4% of total industry): Substantial reductions in energy-based emissions occur from changes in efficiency and the use of alternative fuels in process equipment (e.g., by installing new higher-efficiency kilns). Nevertheless, total reductions are relatively small once process emissions are figured in. Both cement and lime production processes release large quantities of CO_2 when the limestone ingredients are calcined. There are no actions included in the analysis that reduce these emissions.
- *Iron and steel* (reduction of 0.5 Mt CO_2e, 2% of total industry): Emission reductions are small in this subsector, suggesting that opportunities are

costly. Reductions result from switching to electricity, making efficiency improvements in reheat furnaces, and recovering gas from basic oxygen furnaces for use in other processes. The increased penetration of cogeneration is relatively minor.

- *Metal smelting and refining* (reduction of 1 Mt CO_2e, 3% of total industry): In both Ontario and Quebec, firms that produce magnesium have stated that they will eliminate sulphur hexafluorides (SF_6) from their processes before 2010. Because SF_6 has a very high CO_2 equivalence (1 kg SF_6 = 23,900 kg CO_2), the impact is substantial. However, these reductions are included in the reference case. Emission reductions that do occur in the policy simulations are based mainly on the use of more efficient auxiliary systems and the use of different furnace technologies: oxy-fuel burners on furnaces and sulphur combustion furnaces. Certain newly installed equipment, such as the Kivcet lead smelter, requires coal during operation. This requirement reduces opportunities to switch to less carbon-intensive fuels during the time period of the model simulation.
- *Mining* (reduction of 1.2 Mt CO_2e, 4% of total industry): The mining industry makes a strong switch to electricity from fossil fuels by converting equipment to electrical power. At iron ore mines, direct emissions are lowered by reducing heat loss in iron concentration and pelletizing. The potential for efficiency gains in potash processing is small because fuel use is already noncarbon intensive. Nevertheless, some improvements can be made by recovering off-gas heat in potash dryers, improving dewatering methods, and increasing the number of potash crystallization stages. Indirect GHG reductions are made by saving electricity through improved efficiency of auxiliary units and from computerized process optimization.
- *Pulp and paper* (reduction of 2.7 Mt CO_2e, 9% of total industry): Emission reductions in the pulp and paper industry are fairly significant. Key actions include the selection of more efficient process equipment, such as the adoption of high-intensity dryers in paper mills, improvements to boilers, improved process thermal integration in pulp and paper mills, and reduced water use. Wood-waste cogeneration potential is high.
- *Other manufacturing* (reduction of 3.9 Mt CO_2e, 13% of total industry): Reductions in this diverse subsector occur mainly by upgrading older equipment, primarily boilers, with newer, more energy-efficient equipment.

Costs

Table 5.15 shows costs for the industrial sector. Like the electricity sector, coal mining is affected by changes to production levels, though only in response to changes in domestic demand for coal by the electricity sector. Changes in production result in savings in investment, operating, maintenance, and energy costs.

Table 5.15

GHG abatement costs, industry sector

Type of target	Total cost (billions $1995)	Marginal cost ($/t CO_2e)
Sector	34.41	300
National	11.25	120

Note: Total costs shown reflect the net present value of costs from 2000 to 2022.

Table 5.16

Total GHG abatement costs by individual subsector (billions $1995) under the national target and sector target

Industrial subsector	National target	Sector target
Metal smelting and refining	0.58	1.92
Pulp and paper	0.85	3.62
Other manufacturing	2.85	17.16
Mining	0.07	0.79
Iron and steel	-0.02	0.20
Industrial minerals	0.53	0.95
Chemicals	0.03	0.41
Natural gas extraction	2.12	1.44
Petroleum extraction and refining	0.99	3.63
Coal mining	2.55	3.60

The industrial sector bears substantial costs for meeting the reduction target, though the degree differs according to whether reductions are made under a sector or a national target. Industry costs under the national target are less than half of those under the sector target because the high-cost actions are not implemented, and greater reductions occur in other sectors at a lower overall cost to the economy.

Abatement costs to individual subsectors are shown in Table 5.16 for the national target simulation. The distribution of costs between subsectors generally corresponds to that subsector's share of total industrial emission reductions.

The cost pattern changes when industry meets a sector target. In this case high-cost actions across all subsectors are drawn on, in particular those in "other manufacturing." In all runs the pulp and paper, petroleum refining, and industrial minerals subsectors contribute a similar share to total industrial costs. The chemical manufacturing, pulp and paper, mining, and iron and steel industries show significant increases in cost.

Table 5.17

GDP, productivity, input costs, and capital costs, by type of firm in 2010

Industry subsector category	GDP (millions $1986)	Unit cost changes (GDP deflators)	Productivity (total factor productivity)	Capital stock (millions $1986)
Resource-based goods	-3.7	4.1	-5.0	6.5
Energy	-2.7[a]	7.1	-8.7	9.1
Chemicals	-3.2	1.1	0.7	-3.6
Pulp and paper	-1.6	-1.7	-1.5	4.1
Metals	-6.3	3.2	-1.4	0.1
Durable and investment goods	-4.9	6.4	-3.1	0.1
Business-related services	-0.9	4.0	-2.0	3.0
Consumer goods and services	-1.6	7.3	-0.2	-0.1
Government and social services	-0.1	6.7	-0.1	2.1

a Coal mining is impacted by -41%, oil refining -5.8%, and oil and gas mining 0.3%.
Source: Informetrica, *Macroeconomic Impacts of GHG Reduction Options: National and Provincial Effects* (Ottawa: Informetrica, 2000), Tables 155, 157, 158, 159, and 161.

Impacts on Economic Output and Competitiveness

In Chapter 4 we noted that meeting the Kyoto Protocol is expected to have a negative impact on economic output and productivity. These impacts vary significantly by subsector. Expected changes in economic output, productivity, input costs, and capital stock are shown in Table 5.17 for major types of firms and in greater detail for resource-based subsectors. Notable impacts to industry subsectors include the following.

- *Energy production:* The impact of GHG abatement on coal producers is a major part of negative total industry sector impacts: the subsector sees a 41% reduction in GDP in 2010. This impact reflects changes in coal energy demand. Energy production accounts for about 5% to 12% of the impact on GDP.
- *Energy-intensive resource-based subsectors:* These firms are likely to be among the more severely affected in the longer term. These are many of the Tier 1 industries represented within CIMS as distinct submodels whose direct total costs are broken out in Table 5.16. The macroeconomic results suggest that these resource subsectors will account for 20% to 30% of overall reductions in GDP. Many of these industries will be among those supplying materials to meet the additional investment demand and initially are unlikely to be severely affected. However, this initial benefit is offset later by reductions in business investment from adverse effects on industry returns. Industries that benefit most from increased product demand and

from energy-cost savings are affected less negatively. For example, the chemical manufacturing and industrial minerals subsectors see a large reduction in energy costs, a reduction that yields a direct increase in returns. Resource industries are likely to be affected more negatively if other countries do not make changes to their processes to meet GHG reduction targets.

- *Producers of highly manufactured goods/consumer products:* These firms are included in "Other Manufacturing" as defined in CIMS and, as noted earlier, are relatively small energy consumers and GHG producers, but they account for a significant share of economic output. Because foreign trade dominates their markets, they are sensitive to changes in their costs of production and prices relative to foreign competitors. Increases in domestic unit costs of production are likely to produce a consumer shift toward imports of these types of commodities.

Other

This sector is comprised of non-energy-related emissions from agriculture, land use, propellants, hydro-fluorocarbons (HFC), and anaesthetics, which are 80 Mt CO_2e in 2010 in the reference case. Agriculture accounts for almost 90% of emissions from these sources. The inclusion of a forestry sink of 10 Mt CO_2e and an agricultural sink of 6 Mt CO_2e reduces this sector's emissions to 64 Mt CO_2e.

In addition to these sinks, agriculture and forestry measures were defined and included in this analysis. As mentioned in Chapter 3, they were not modelled explicitly in CIMS but were calculated externally to the model. Thus, their impact does not change between the national and the sector targets. Their impact is also highly uncertain because GHG estimation in this area is complex and new.

Table 5.18

GHG emission reductions, agriculture measures

Agriculture measure	Reduction in 2010 (Mt CO_2e)
Improved nutrient management	0.9
Increase no-till	2.1
Decrease utilization of summer fallow	1.7
Increase permanent cover program (high cattle increase)	0.1
Grazing strategies	2.6
Combined feeding strategies	0.6
Expand shelterbelts on Prairies	0.02

Agricultural Measures

About 10% of Canada's GHG emissions are attributed to agricultural activities. Methane and nitrous oxide account for almost all of these emissions through sources such as livestock digestion, manure management, agricultural soil activities, and agricultural residue burning. Irrigation and tillage practices may also contribute GHGs. This estimate does not include emissions from fertilizer production, which are attributed instead to the industrial sector. Actions that target agricultural GHG emissions are shown in Table 5.18.

Forestry Measures

These measures relate to land use and forestry actions that increase forestry sinks and reduce forestry sources of GHG emissions. The actions included in our analysis are all afforestation actions that differ by tree species, the type of afforestation, and region. These actions are summarized in Table 5.19. Afforestation actions are characterized by large up-front costs and carbon sequestration benefits that are initially low but become substantial after several decades.

Table 5.19

Summary of GHG reduction potential, afforestation actions

Action	Total planting (ha)	Carbon sequestration ($Mt\ CO_2e$) 2010	2000-50
Fast-growing plantations	50,000	1.31	n/a
Prairie shelterbelts[a]	169,000	0.15	29.0
BC block plantations	169,000	0.04	35.2
Prairie block plantations	260,000	0.37	71.4
Eastern block plantations	195,000	0.22	68.6
Total	843,000	2.09	

a Prairie shelterbelts are planted for the purposes of soil conservation and windbreaks. They are normally planted on lands that have been exclusively agricultural.

Notes: Only above- and below-ground tree biomass carbon is included in the net sequestration estimates for the fast-growing plantation action and the Prairie and BC actions. The Eastern Canada actions also include soil and nontree biomass carbon. Emissions from the use of fossil fuels in planting are accounted for in the estimates. For the fast-growing species action, the assumption is that harvesting, if it occurs, will happen at ages 13-15 years, and the area is replanted. Over the 2000-50 period, the net carbon sequestration will then depend on how harvesting of afforested areas and carbon in the resulting forest products are treated in the Kyoto Protocol.

Source: Based on information in Forest Sector Table and Canadian Forest Service of Natural Resources Canada and Environment Canada, *Options for the Forest Sector to Contribute to Canada's National Implementation Strategy for the Kyoto Protocol* (Ottawa: National Climate Change Process, 1999).

Conclusion

The actions that occur in each sector so that Canada reaches its reduction target are in most cases simply a stronger application of technologies that already exist. Certain sectors are significantly greater contributors to reductions than others and consequently bear greater costs. This difference is based directly on the dynamics and structures of the sectors. Transportation is a large source of reductions through changes to transportation modes, improved vehicle efficiencies, and reductions in travel demand. The electricity sector also plays a large role, particularly through the emergent technology of CO_2 capture and sequestration rather than through renewable technologies such as wind and photovoltaic. The industrial sector is a strong contributor, though more by virtue of being a large source of emissions rather than through the availability of cost-effective actions. These actions are more limited than in other sectors because firms have often already pursued cost-effective options and are more physically limited in their ability to reduce energy use. Most reductions occur from actions that are specific to certain industries rather than through generic actions that are applicable to all. Of these reductions, those made in the natural gas extraction and petroleum-refining subsectors are the most significant. The residential sector and the commercial/institutional sector are smaller contributors to overall reductions, mainly because they are less significant sources of emissions. Still, many actions taken in those sectors, particularly in improving heating efficiency and controlling landfill gas emissions, can yield significant reductions.

The differences in reductions and costs by sector have important implications for how cost incidence is shared regionally. Sectors and subcomponents of those sectors are grouped geographically. For instance, natural gas and petroleum extraction dominate in the West; regions have very different ways of generating electricity; and commercial and institutional buildings are concentrated in Ontario and Quebec. This dimension is explored in the next chapter.

6
Regional Estimates

In announcing the National Climate Change Process, Canada's premiers and the federal government agreed that no region should bear an unreasonable burden from implementing the Kyoto Protocol. However, in the absence of careful policy design, and perhaps compensating mechanisms, it is clear that the costs of GHG emission reduction will vary not only by sector but also by region. In this chapter we examine the regional impacts of the policy simulations. We first review the relative contributions of regions to meeting Canada's Kyoto target and then present the results for each of the seven regions that were modelled separately.

As in Chapters 4 and 5, we focus on two contrasting policy simulations (or paths): the sector target (Path 1) and the national target (Path 4). While we do not focus here on policy implications, the results from this chapter inform the policy discussion of Chapter 7. The total costs presented here are the net present value of costs from 2000 to 2022. Annual per capita and household costs are based on projected household figures in 2010. These costs are calculated based on equal annual payments over 22 years.

Regional Comparison

Canada's regions differ in terms of their resources, economies, and population sizes. Thus, the costs described in Chapter 4 are distributed unevenly. A significant share of emission reductions occurs in two provinces, Alberta and Ontario, while per capita costs are borne most heavily by Alberta and Saskatchewan. Figure 6.1 shows where emission reductions are made, and Figure 6.2 shows the distribution of total costs.[1] A detailed summary of emission reductions and costs is presented in Table 6.1. This information shows that some regions' costs are not proportional to their share of Canada's population and that costs in certain regions are sensitive to how the reduction target is reached.

In both policy simulations, reductions configure themselves in a least-cost pattern across regions, either within a sector (under the sector target) or

Figure 6.1

Contribution of each region to emission reductions in 2010

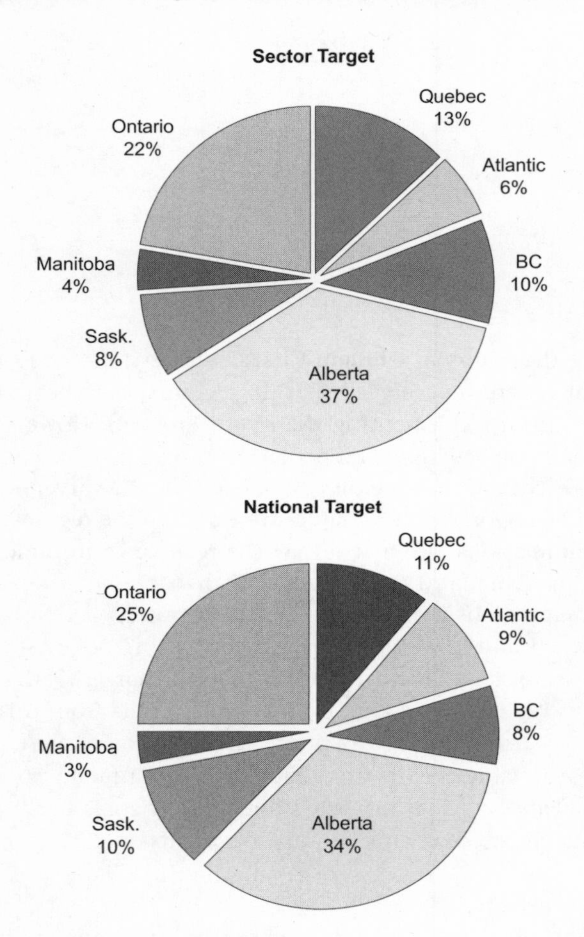

across the economy as a whole (under the national target). Alberta's carbon-intensive economy – which has high GHG intensity in all sectors – offers some reductions at lower costs than regions with less carbon-intensive economies. Alberta's coal-based electricity emissions can be reduced through various means, and its industrial sector can provide more reductions at lower costs than elsewhere. Further insight into each region's ability to make reductions is provided later in this chapter.

Total costs are greatest where most of the population and economic output is based: Ontario and Quebec. This is particularly true because emission

Figure 6.2

Contribution of each region to direct total costs

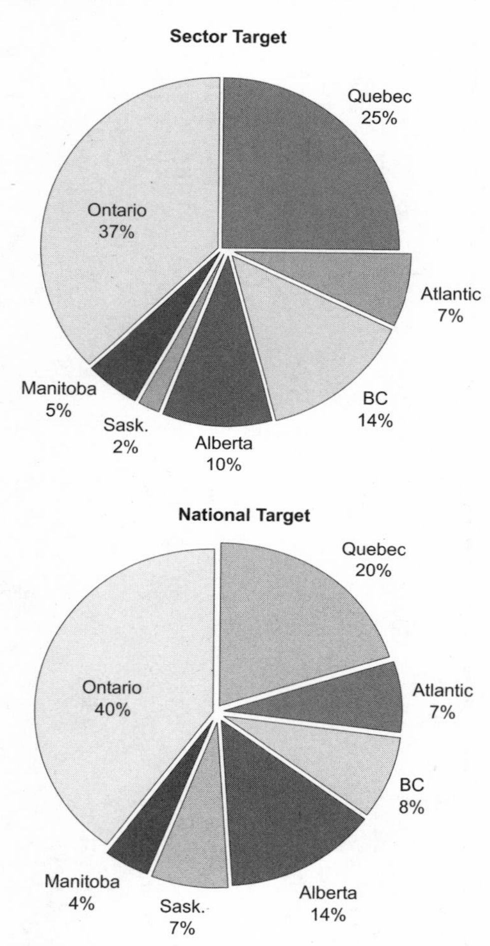

reductions from transportation are important to overall reductions, and regions do not have markedly different reduction options in this sector. Thus, similar degrees of emission reduction are made at similar costs, but because of Ontario's and Quebec's population sizes, their total transportation reduction costs are high. The same is generally true for the residential and commercial/institutional sectors, although these reductions are less important to total costs. The industrial sector is technologically and spatially heterogeneous, so costs are affected by the regional makeup of industry. Because industrial output is based most heavily in Ontario and Quebec,

Table 6.1

Summary table of emission reductions and costs by region

		BC	Alberta	SK	Manitoba	Ontario	Quebec	Atlantic	**Canada**
Reference case	Total GHG emissions (Mt CO_2e in 2010)	74.1	233.2	65.2	25.1	213.1	96.2	53.1	**760.0**
	GHG intensity (t CO_2e/capita in 2010)	16.7	70.3	56.4	19.3	16.2	11.8	21.1	**21.8**
	Emissions growth (1990-2010)	47%	40%	42%	32%	17%	12%	16%	**27%**
Sector target	GHG reductions (Mt CO_2e in 2010)	15.0	54.2	12.7	6.2	36.6	19.9	8.9	**153.5**
	Degree of reduction (% reduced)	20%	23%	19%	25%	16%	21%	17%	**20%**
	Total cost (billions $1995)	8.42	5.89	1.10	2.80	23.04	15.63	4.28	**61.20**
	Per capita cost ($1995/person)	1,871	1,775	912	2,156	1,756	1,924	1,700	**1,796**
National target	GHG reductions (Mt CO_2e in 2010)	14.6	62.1	18.9	5.6	48.1	18.6	16.4	**184.3**
	Degree of reduction (% reduced)	20%	27%	29%	22%	22%	20%	31%	**24%**
	Total cost (billions $1995)	3.53	6.39	3.05	1.9	18.01	8.74	3.07	**44.7**
	Per capita cost ($1995/person)	784	1,926	2,529	1,463	1,373	1,076	1,219	**1,311**

Note: Total costs and per capita costs represent the net present value of direct costs from 2000 to 2022. Totals for Canada do not match exactly the totals in Table 4.2 and Figure 4.6 due to rounding.

small reductions by individual industries translate into sizeable provincial costs. In general, the relationship between population size and costs can be seen in Table 6.1 by comparing "Degree of Reduction," "Total Cost," and "GHG Reduction." Neither Ontario nor Quebec has the largest degree of emission reductions.

A closer inspection of Table 6.1 reveals the differences in how the two reduction strategies affect regions relative to their population sizes. Under the sector target, per capita costs are higher than the national average in British Columbia, Manitoba, and Quebec, while Saskatchewan has substantially lower per capita costs, and Alberta and Ontario are about even. The greatest costs under this policy are in the industry and transportation sectors, which are pushed to make costly reductions to meet their sector reduction targets.[2] When the sector target restrictions are lifted, and more emissions are reduced in the electricity sector (and to a lesser degree in the residential and commercial/institutional sectors), the provinces with lower-cost reduction potentials in that sector – principally Alberta and Saskatchewan – bear a greater share of national costs.

Actions taken in each sector have different regional implications, which significantly impact regional costs. We note a few below.

- *Electricity:* Where electricity is more fossil fuel based, such as Alberta, Saskatchewan, Ontario, and the Atlantic region, the electricity sector provides substantial emission reductions. The greatest reductions occur in Alberta and Saskatchewan, where emissions are reduced by 75% in 2010 in the national target. In the region-specific sections below, changes in electricity costs are not provided because they emerge instead in the demand sectors through changes in electricity prices.
- *Transportation:* This sector is important to reductions in all regions, with a similar per capita amount of reduction occurring in each. Total reductions and costs by region match the relative population size, but some actions figure more prominently in certain areas. Transit improvements occur mostly in regions with large urban centres, such as Ontario and Quebec; costs are somewhat greater in regions with these improvements.
- *Residential and commercial/institutional:* These sectors are not large contributors to direct emission reductions in any region, but they are nonetheless important, because several actions are relatively low cost or even beneficial. For example, significant benefits can occur from efficiency improvements and the application of municipal actions, including savings in community infrastructure costs.[3] As in other demand sectors, the costs to residential and commercial buildings also reflect changes in their electricity costs, which rise due to increased electricity prices (remember that the electricity sector's costs are passed on in price). This cost element can be significant because buildings are major users of electricity (see Figure

5.18). Although residential and commercial building types are fairly similar across regions, changes in relative energy prices affect whether an action is undertaken. The degree of reduction by region is therefore quite different.

- *Industry:* These emission reductions vary according to each province's industrial structure. Industry emissions and emission reductions are greatest in Alberta, Saskatchewan, and Ontario. As in the residential and commercial/institutional sectors, changes in electricity costs are included in this sector's costs.

In the following region-specific sections, we discuss cost impacts more fully, including sectoral, macroeconomic, and household costs, energy price impacts, and sources of reduction.

British Columbia

British Columbia is not a GHG-intensive region, accounting for only about 10% of Canada's emissions (which is less than its share of population). At 16 t CO_2e in 1996, its per capita emissions are the lowest in Canada after Quebec. This low amount is due to significant hydroelectric generation and to relatively low industrial emissions. British Columbia's most important industry, pulp and paper, uses considerable wood waste, a non-GHG-emitting fuel. Other important industry subsectors – mining and "Other Manufacturing" – do not emit a lot of GHGs. The majority of emissions in industry come from natural gas extraction and transmission, petroleum refining, and metals refining.

Figure 6.3 shows emissions growth by sector in the reference case. Total emissions in 1997 were 63 Mt CO_2e. Transportation is the single biggest source, followed by industry and the commercial/institutional sector. Population and economic growth will increase the demand for energy, which will be met by oil and natural gas. In the absence of further action, provincial emissions are forecasted to be 47% higher in 2010 than in 1990. This growth rate is exceeded only by that in Alberta and Saskatchewan.

In transportation, industry, and electricity, emissions are growing faster than population. Transportation emissions are estimated to grow 50% between 1990 and 2010, while population is expected to grow only 33%. British Columbia's per capita transportation emissions are higher than those of Ontario and Quebec, where urban form and availability of transit have lowered car use for commuting. According to a survey of commuting behaviour, Vancouver had the highest average commuting time and one of the lowest rates of public transit use and walking in Canadian urban areas.[4]

BC industry emissions are expected to grow fairly rapidly, fuelled principally by rapid growth in natural gas extraction in the northeast. Emissions are also expected to grow in chemical manufacturing, although this is a

Figure 6.3

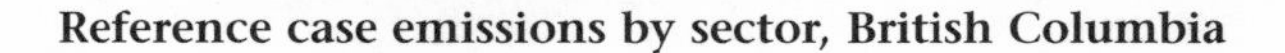

Reference case emissions by sector, British Columbia

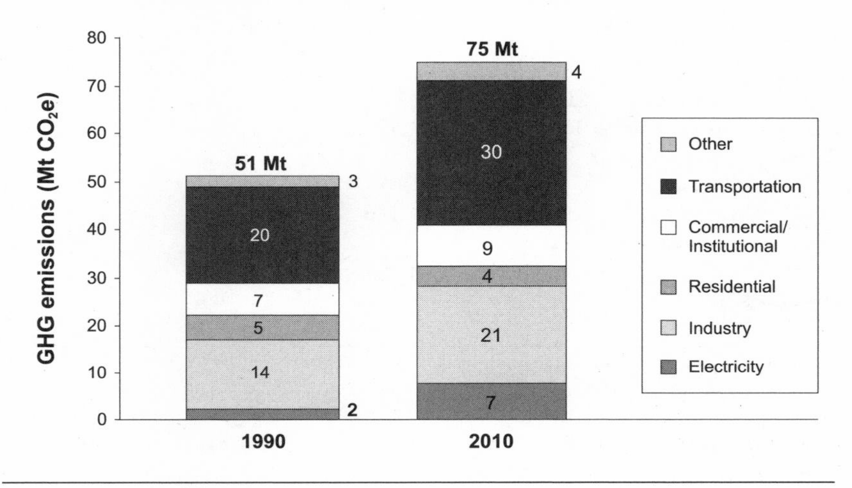

small share of total industry emissions. Other subsectors are expected to have much more sluggish growth. Overall, industry emissions are 48% higher in 2010 than in 1990.

Electricity emissions are also increasing rapidly as gas-fired generation meets a growing demand for electricity; emissions are projected to more than double from 1990 to 2010. No further large hydro projects are expected, though certain sites will be redeveloped.

GHG Policy Impacts

Emission reductions are relatively limited in British Columbia because the structure of its economy makes actions more costly than those of other provinces. As shown in Table 6.2, emissions are reduced by approximately 20%, one of the lowest reductions in any region, at costs of $3-$8 billion. The impact on GDP is relatively low, estimated to be -1.2% in 2010. This low impact is due mainly to limited costs in the forestry and pulp and paper industry subsectors.

Figure 6.4 shows the sources of emission reductions for each strategy. British Columbia's potential to reduce emissions is greatest in the transportation sector and to a lesser extent in the industrial sector. Total reductions are slightly greater in the sector target simulation because transportation and industry are pushed further by their targets. Under the national target, emission reductions are instead made in the electricity sector of regions where there are significantly lower-cost opportunities.

Table 6.2

Emission reductions and costs, British Columbia

	Reference case	Sector target simulation	National target simulation
Emissions in 2010 (Mt CO_2e)	74.1	59.1	59.5
Energy consumed in 2010 (PJ)	1,194	989	985
GHG intensity in 2010 (t CO_2e/person)	16.7	13.1	13.4
Emissions relative to 1990[a] (%)	145	116	117
Emissions reduced in 2010 (Mt CO_2e)	n/a	15.0	14.6
Total cost (billion $)	n/a	8.4	3.5
GPD impact (%)	n/a	n/a	-1.2

a 1990 GHG emissions in British Columbia are 51 Mt CO_2e.

British Columbia's hydro-dominated generation offers little room for GHG reductions. Actions in the province are generally more expensive than those taken elsewhere to the extent that switching to small hydropower from natural gas is considered to be more costly than switching from coal to natural gas (which is a possibility in many other regions).

Transportation is a far greater contributor to reductions than electricity supply. Under the national target, transportation emissions are 22% less than those of the reference case, and GHG intensity is improved to 5.2 t CO_2e per person. Improvements to transit make up 21% of the total reductions in this sector. Speed-control measures, a major source of reductions nationally, are less effective in this province because travelling speeds are already lower; nevertheless, they contribute about 15% to total transportation emission reductions. Ten percent of reductions are attributable to vehicle efficiency standards. Transportation emission reductions cost $2.3 billion (sector target) and $1.2 billion (national target), about one-third of the total costs in British Columbia. These costs are equivalent to $104 per household per year (national target) and $205 per household per year (sector target).

Industry costs are more than double those of transportation under the sector target, yet they contribute far fewer reductions. High-cost actions are abandoned under a national target, so costs decline to $730 million from $5 billion. Nevertheless, while costs decline more than 80%, emission reductions decline by only 8%. Of the industry sector reductions that remain, about 1 Mt CO_2e occur in natural gas extraction. Pulp and paper mills contribute the next largest share, 0.4 Mt CO_2e. This subsector's emissions already decline from 1990 levels in the reference case due mainly to high wood-waste and electricity use. Despite this already significant reduction,

Figure 6.4

Contribution of sectors to emission reduction in British Columbia, 2010

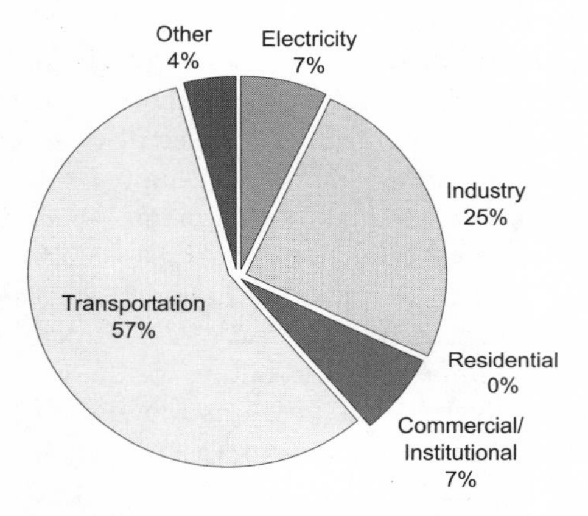

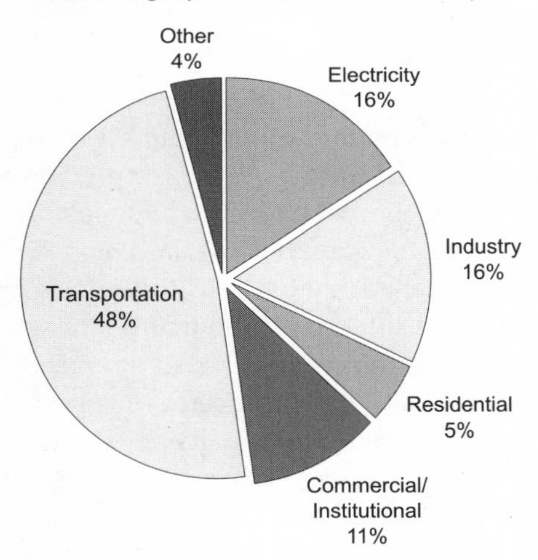

additional reductions occur from technology improvements (in particular by using more efficient process equipment in chemical pulping) and through further use of wood waste. Other manufacturing, which in British Columbia is comprised mainly of wood products and food processing, also makes some reductions. Of the remaining industries, little occurs in metal refining because certain newly installed equipment, such as the Kivcet lead smelter, requires coal during operation, reducing opportunities to switch to less carbon-intensive fuels in the simulation period. Both coal and metal mining increase their use of electricity. Unlike other coal regions where emissions decline from reduced production (due to changes in the electricity industry), British Columbia's coal production is not affected; it is predominantly metallurgical coal for export.

The residential and commercial/institutional sectors contribute comparatively less to total emission reductions: a total of 16% under the national target. However, because opportunities for electricity and industrial emission reductions are more limited in British Columbia, these sectors account for a higher percentage of the total than in many other regions. Although the residential sector's total reductions are small – 20 Kt CO_2e under the sector target and 700 Kt CO_2e under the national target – costs to households increase as their electricity costs rise.[5] Total residential sector costs in the province are $1.1 billion and $1.6 billion for the sector target and the national target respectively. These amounts roughly translate into yearly household costs of $102 and $147. In contrast, the commercial/institutional sector sees cost savings of $150 million under the sector target and $90 million under the national target, with reductions of 1.1 Mt CO_2e and 1.5 Mt CO_2e, respectively.

Alberta

Fossil fuels dominate the Alberta economy: almost all electricity is generated by coal, and its industrial sector is largely centred on fossil fuel extraction and processing. With the largest livestock population in Canada, the dependence on fossil fuels for space heating, and high vehicle use, it is not surprising that Alberta is Canada's most GHG-intensive region. At 71 t CO_2e per capita in 1996, its GHG intensity is three times those of British Columbia, Ontario, and Manitoba and five times that of Quebec. While accounting for a tenth of Canada's population, Alberta contributed 199 Mt CO_2e in 1996, close to a third of Canada's total GHG emissions. Industry is by far the biggest contributor to this total, followed by electricity and transportation. Emissions from agriculture (which are mainly methane from raising livestock and lost nitrogen oxides from soil) contribute approximately three times as many emissions as the direct emissions from Alberta's residential sector.

Figure 6.5 shows Alberta's emissions and their growth in the reference case. Major growth occurs in industry, transportation, and electricity generation. Industrial emissions grow 60% between 1990 and 2010, principally because natural gas production increases to satisfy US demand. Oil sands production also increases with the expansion of existing facilities and the development of new products. Efficiency is projected to improve greatly in these industries; however, with 80% of Canada's crude oil reserves, 90% of its

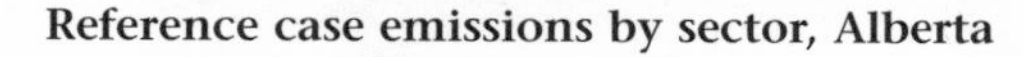

Figure 6.5

Reference case emissions by sector, Alberta

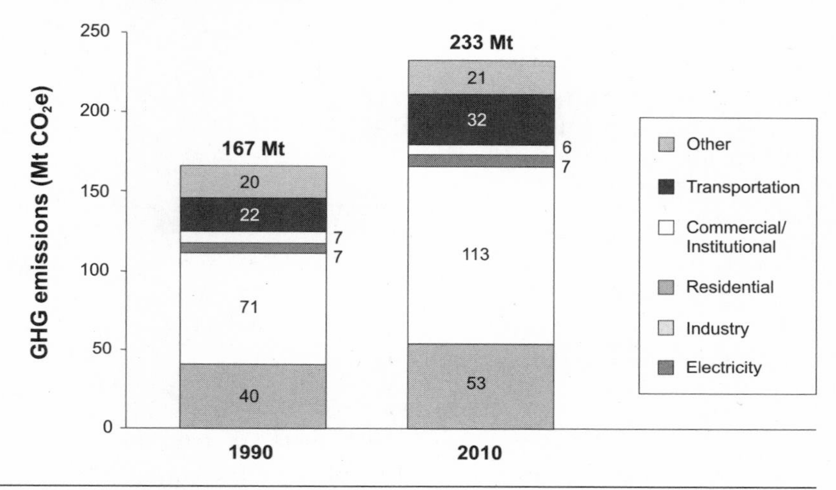

natural gas, and all of its bitumen and oil sands reserves, it is not surprising that Alberta sees such strong emission increases.

After industry emissions, transportation emissions grow the most. With an increase of 45% between 1990 and 2010, emissions grow considerably faster than population. Per capita emissions in this sector are already the highest in Canada – GHG intensity per capita is about twice that of Ontario and Quebec. The number of cars on the road has doubled in the past 20 years, and the number of persons per vehicle is the lowest in Canada. This trend has been influenced by factors such as urban sprawl, the mix of local economic activities, and fuel tax rates.[6]

Electricity emissions grow 34% between 1990 and 2010. This increase is not substantially more than population growth. New electricity demand is expected to be met by combined cycle natural gas units; existing coal capacity will be maintained as retiring coal units are replaced with more efficient units. System efficiency improves in the reference case from 33% in 1990 to 39% in 2010. Albertans use very little electricity as a share of total energy demand – about 10%, compared with 20%-50% in most other regions. Abundant and accessible fossil fuel supplies have favoured the use of natural gas for heating in homes and commercial buildings, and for industrial use.

GHG Policy Impacts

In both policy simulations, Alberta contributes the most emission reductions

Table 6.3

Emission reductions and costs, Alberta

	Reference case	Sector target simulation	National target simulation
Emissions in 2010 (Mt CO_2e)	233.2	179.0	171.1
Energy consumed in 2010 (PJ)	2,427	1,954	2,025
GHG intensity in 2010 (t CO_2e/person)	70.3	54.0	51.6
Emissions relative to 1990[a] (%)	140	107	102
Emissions reduced in 2010 (Mt CO_2e)	n/a	54.2	62.1
Total cost (billion $)	n/a	5.9	6.4
GPD impact (%)	n/a	n/a	-1.9

a 1990 GHG emissions in Alberta are 167 Mt CO_2e.

to meet Canada's reduction target because there are many relatively low-cost opportunities to switch from GHG-intensive fuels, and because CO_2 capture and sequestration is projected to be less expensive than many industry and transportation actions elsewhere. Table 6.3 summarizes emission reductions and costs. Alberta's costs of $5.9 and $6.4 billion (for sector and national targets respectively) are among the highest in Canada. The large weight of energy-related industry in Alberta's economy translates into a relatively strong negative impact on total GDP, since these industries are a source of significant GHG reductions (and direct costs).

Alberta's potential to reduce GHG emissions is greatest in the electricity sector and to a lesser extent in the industrial sector (Figure 6.6). Transportation also makes considerable reductions. Total emission reductions are greater under the national target because the country turns to Alberta's electricity sector for reduction opportunities when high-cost sectors (transportation and industrial sectors) do not have to meet a sector target.

About 60% of Alberta's reductions occur in the electricity sector as some generation shifts from coal to natural gas and as some CO_2 emissions from remaining coal generation are captured and stored underground. The latter operation occurs predominantly under the national target. Electricity GHG intensity per person drops dramatically from 16 to 4 t CO_2e in the national simulation. Electricity prices increase by 85%, causing significant reductions in electricity demand. Lower electricity production accounts for approximately a quarter of Alberta's GDP impact.

The industrial sector supplies the next largest share of reductions although under the policy simulations its GHG emissions are 44% above 1990 levels and its emissions are reduced by only 10%. These industrial reductions are among the most cost effective in Canadian industry. This sector reduces 11

Figure 6.6

Contribution of sectors to emission reduction in Alberta, 2010

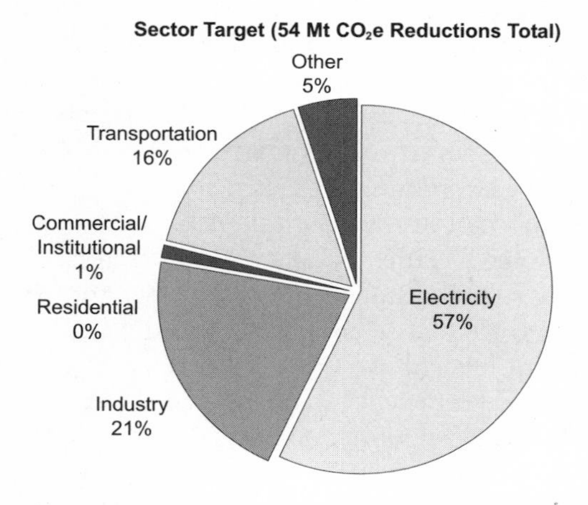

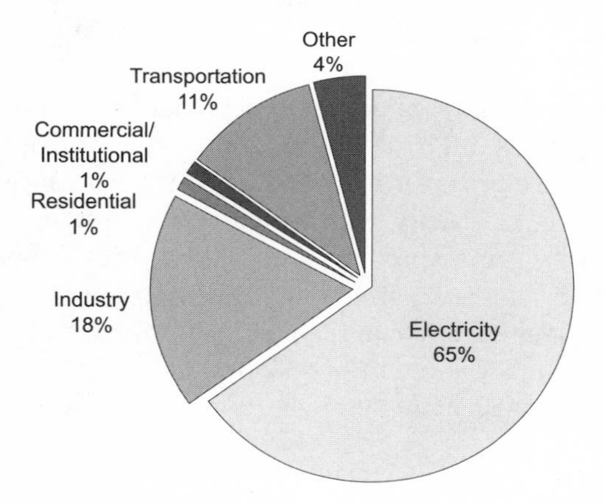

Mt CO_2e at a cost of $1.6 billion in the national target simulation. An additional 700 Kt CO_2e of reductions in the sector target simulation cost about as much as the first 11 Mt CO_2e, which suggests that actions that cost more than $120/t CO_2e provide few reductions. Actions in the petroleum extraction and refining subsector (7.9 Mt CO_2e) and to a lesser extent in natural gas extraction (2.3 Mt CO_2e) are responsible for most industrial sector reduction. The coal-mining subsector contributes the next largest share, 0.3 Mt CO_2e, though most of this amount occurs from reduced production because the electricity sector demands less coal. Other reductions are limited since little processing occurs and its open-pit operations do not allow for capture and use of methane. Some other emission reductions occur in industrial minerals and "Other Manufacturing." Of the remaining industries, direct emissions in pulp and paper production and chemical manufacturing increase because these subsectors take advantage of cogeneration, which is attractive because of the electricity price increases. Cogeneration is also more widely adopted in Alberta's petroleum refineries, which, like others across the country, become net producers of electricity.

The transportation sector shows significant reductions, though at a smaller scale than those of the electricity and industrial sectors. In the national target simulation, emissions are cut by about 21%, and per capita GHG intensity improves to 7.6 t CO_2e per person. Speed-control actions are the largest contributor to reductions (19%). Improvements to transit provide another 18%, and vehicle efficiency standards add another 12%. Total emission reductions cost $2.3 billion (sector target) and $1.4 billion (national target). These reductions are equivalent to $177 per household per year (national target) and $289 per household per year (sector target).

The residential and commercial/institutional sectors contribute little to the province's emission reductions: only 3% under the national target. Nevertheless, at 850 Kt CO_2e (residential) and 870 Kt CO_2e (commercial/institutional), reductions are only slightly less than those in British Columbia. Costs are relatively high – $980 million in the residential sector and $1.7 billion in the commercial/institutional sector – because these sectors use a fair amount of electricity, which increases sharply in price. The residential costs translate into approximately $122 a year per household.

Agricultural and afforestation actions contribute more emission reductions than the residential and commercial/institutional sectors. Reductions of 2.5 Mt CO_2e are made at a cost of $680 million, primarily through grazing and grassland management and by increasing the area that is not tilled.

Saskatchewan

In spite of its low population, Saskatchewan emits substantial GHGs. With one-third the population of British Columbia, it is responsible for close to the same amount of emissions, 59 Mt CO_2e in 1996. Like Alberta, Saskatchewan depends on coal to generate electricity and has a growing fossil fuel industry. Its transportation and residential emissions are also high on a per capita basis, thus making its total GHG intensity 57.8 t CO_2e/person in 1996, second only to that of Alberta. As shown in Figure 6.7, industry, agriculture, electricity, and transportation contribute almost equally to Saskatchewan's total emissions in 1990 (agriculture is "Other" in the figure).

Emissions in all sectors except for residential grow considerably. Saskatchewan shows the third highest growth in total emissions next to British

Figure 6.7

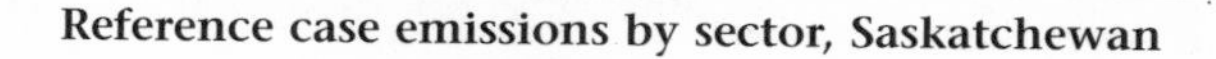

Reference case emissions by sector, Saskatchewan

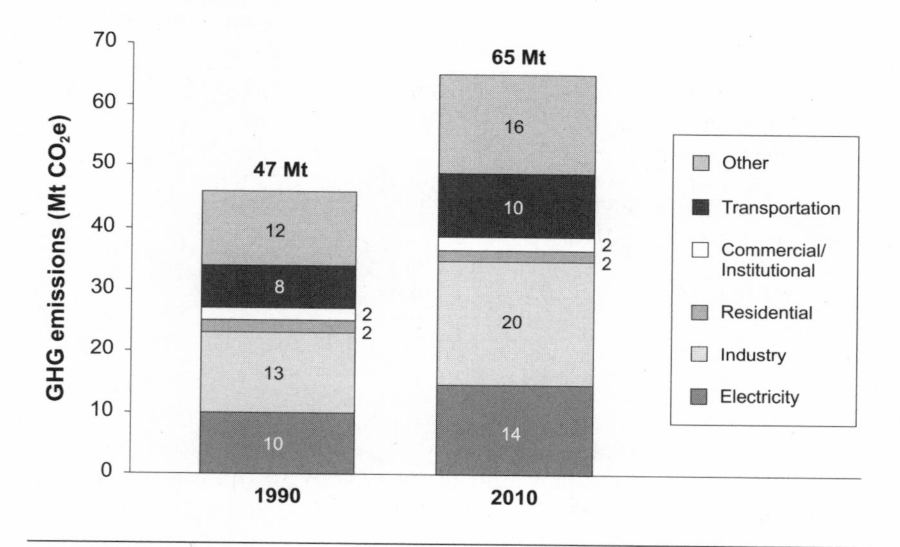

Columbia and Alberta; however, on a per capita basis, Saskatchewan shows the highest rate of increase. Growth is particularly strong in industry and electricity, at 56% and 43% respectively. Industry emissions are mostly attributable to natural gas extraction and petroleum refining but also stem from mining, in particular potash, which requires significant quantities of heat during processing. Saskatchewan's natural gas production is expected to rise to meet US export demand. Transportation becomes more GHG-intensive between 1990 and 2010, from 7.9 to 8.9 t CO_2e/person. Electricity growth is similar to that of Alberta as new demand is met by natural gas, while existing coal capacity is maintained.

GHG Policy Impacts

Reductions in Saskatchewan are significant (between 20%-30%) because, as in Alberta, there are relatively low-cost opportunities in the electricity and industrial sectors. Direct cost and GDP impacts are also high (see Table 6.4). Emission reductions are made at a cost of $1.1-$3.1 billion. By 2010 GDP is expected to decline by 2.5%, which is higher than Alberta's impact. Total reductions are slightly greater in the national target simulation because, as in Alberta, emission reductions are found in Saskatchewan's electricity sector when transportation and industry sectors make fewer reductions. Figure 6.8 shows the sources of reductions by sector.

Table 6.4

Emission reductions and costs, Saskatchewan

	Reference case	Sector target simulation	National target simulation
Emissions in 2010 (Mt CO_2e)	65.2	52.5	46.3
Energy consumed in 2010 (PJ)	596	502	486
GHG intensity in 2010 (t CO_2e/person)	56.4	45.4	40.1
Emissions relative to 1990[a] (%)	139	112	99
Emissions reduced in 2010 (Mt CO_2e)	n/a	12.7	18.9
Total cost (billion $)	n/a	1.1	3.1
GPD impact (%)	n/a	n/a	-2.5

a 1990 GHG emissions in Saskatchewan are 47 Mt CO_2e.

Under the national target, the electricity sector provides the majority of GHG savings and contributes most to costs. As in Alberta, shifts from coal to combined cycle natural gas generation and the capture and sequestration of CO_2 provide the bulk of these emission reductions. Under the national target, Saskatchewan has one of the highest electricity price increases – 68% – next to Alberta.

Saskatchewan's growing fossil fuel extraction industry outweighs any significant decline in industrial emissions under the policy simulations. Nevertheless, due to the large amount of industry emissions, this sector is the next largest source of reductions. Similar industry costs occur under both policies – approximately 2.5 Mt CO_2e are reduced at a cost of about $1.4 billion. Most industry GHG reductions occur in the natural gas subsector. The degree of emission reductions is comparable to that of Alberta but occurs at less expense since reductions are mostly made to transmission rather than extraction and processing. Not only is Saskatchewan's production increasing quickly, but also all of the gas from Alberta that is transported east passes through this province.

The petroleum industry contributes the next largest share, 410 Kt CO_2e, to total industry reductions. While little growth is expected in petroleum refining, significant growth in extraction is expected. Control of fugitive emissions figures significantly in reductions. In petroleum refining, efficiency improvements to steam and heat are made, including the adoption of greater cogeneration, which becomes attractive due to the higher electricity price under the national target. Of the remaining industries, mining, coal mining, and "Other Manufacturing" contribute a handful of reductions (4% of the total). Demand changes are responsible for almost all of coal

Figure 6.8

Contribution of sectors to emission reduction in Saskatchewan, 2010

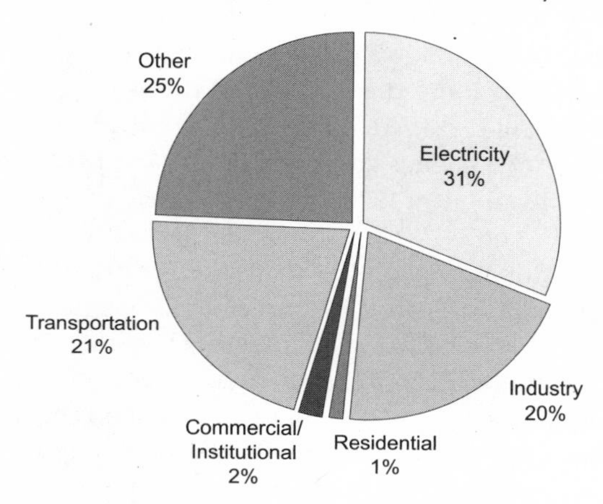

National Target (19 Mt CO_2e Reduction Total)

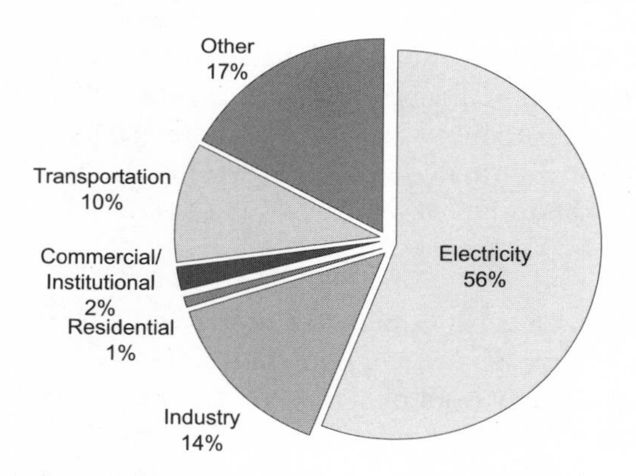

mining's emission reductions; few technological opportunities exist because the coal is not processed. Potash production offers few reductions because the potential for efficiency gains and fuel switching is small.

Transportation emissions in the national target are 19% less than in the reference case, improving GHG intensity to 7.2 t CO_2e per person. But with Saskatchewan's low population, these efficiency improvements do not translate into significant reductions. Because the population is not as concentrated in urban areas, improvements to transit account for only 11% of the sector's reductions (Quebec's and Ontario's are about 25%). The most significant actions are speed control and efficiency standards, which account for 18% and 13% respectively. Emission reductions cost a total of $480 million (sector target) and $420 million (national target), which are equivalent to $158 and $139 per household per year.

The commercial/institutional sector does not increase its direct emissions in the reference case by more than 10% over the period and contributes only slightly to total emission reductions (1.6%) under the national target. Reductions are made at a savings under the sector target ($200 million) and a cost under the national target ($790 million). The residential sector yields even fewer reductions, only 100 Kt CO_2e under the sector target and 170 Kt CO_2e under the national target.[7] The first case provides a net benefit to households – about $27 a year. However, households pay in the second situation (at a cost of about $301 a year) mainly because of electricity cost increases as well as some GHG reduction actions. For both sectors, reductions are less in Saskatchewan than in many other regions because little fuel switching can occur – natural gas has penetrated strongly already. Fuel switching to electricity is not attractive because of the high GHG content of the province's electricity.

Saskatchewan has the greatest capacity to reduce emissions from agricultural actions of any region in Canada. Reductions of 3.2 Mt CO_2e are made at a benefit of $490 million. Increasing nontilled land and decreasing summer fallow are the largest sources of these reductions; grazing and grassland management do not contribute as much as in Alberta. The benefits stem mainly from no-till and summer fallow land practices.[8]

Manitoba

Manitoba's share of Canada's population is 4%, while its share of emissions is 3%. Total emissions in 1996 were only 23 Mt CO_2e. Manitoba obtains most of its electricity from hydropower. In contrast to other regions, agriculture in Manitoba is the greatest source of emissions, followed by transportation. Agricultural emissions mainly occur from the release of nitrogen oxides from agricultural soil. Industry emissions are a distant third, contributing half that of the transportation sector. Economically, Manitoba's service and agricultural sectors are more important than its industry. The industry that does exist is not particularly energy or emissions intensive: a diverse range of manufacturing firms can be found, such as food and transportation manufacturing, clothing, and textiles. More energy-intensive activities include metal mining and smelting, principally nickel, copper, and zinc. Mining production is forecast to drop considerably, and emissions are 50% lower in 2010 than in 1990. Oil and gas pipelines that run through the province release almost as much emissions as all of the province's industry.

Figure 6.9

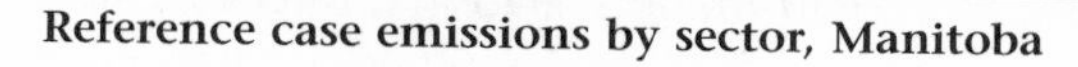

Reference case emissions by sector, Manitoba

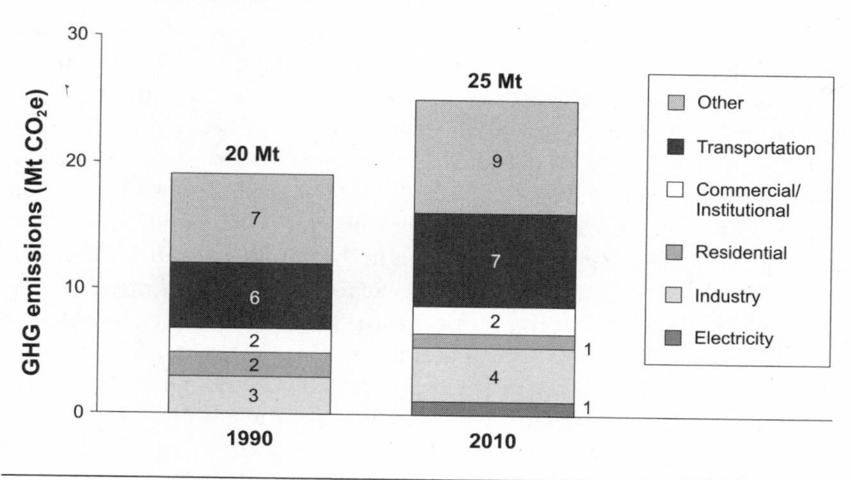

Figure 6.9 shows the share of emissions by sector and their growth in the reference case. As in other regions, transportation emissions grow strongly. Electricity demand growth is met by hydro and some natural gas combined cycle generation. Thus, electricity emissions are expected to increase, though only to 0.39 t CO_2e per person in 2010; Manitoba will not digress from its hydro-dominated path. Greater volumes of natural gas transmission are largely responsible for the increase in industrial emissions.

GHG Policy Impacts

Compared with the economies of the other Prairie provinces, Manitoba's economy is relatively unaffected by the simulated policies. As shown in Table 6.5, under the national target its GDP will be affected negatively by 1.4%, and its direct costs will be $1.9 billion, which are the lowest in Canada. Total emission reductions are few, even though emissions are cut by approximately a quarter, because emissions are negligible to begin with. Thus Manitoba contributes only 3% to Canada's total reductions, mainly from transportation, afforestation, and agricultural actions. Reductions are slightly greater under the sector target because transportation and industry, Manitoba's main sources of emissions, are pushed to make more reductions. Figure 6.10 shows the reductions by sector.

Transportation emissions are 21% less than the reference case under the national target, reducing GHG intensity to 4.5 t CO_2e per person. Improvements to public transit contribute 20%, speed-control measures 17%, and efficiency standards 13%. Transportation emission reductions cost a total of $950 million (sector target) and $630 million (national target). These amounts are equivalent to annual costs of $302 and $199 per household.

Agricultural and afforestation actions account for the next largest share of reductions: 1.7 Mt CO_2e are reduced at a net benefit of $80 million. Reductions occur from a variety of actions, notably grazing and grassland management, and by increasing the area that is not tilled.

Industry reduces the next largest amount of emissions, although reductions amount to only 1.3 Mt CO_2e and 1.8 Mt CO_2e at a cost of $750 million and $2.2 billion (the higher reductions and costs occur under the sector target). The majority of reductions occur in the natural gas subsector, which in Manitoba covers only transmission. Reductions occur mostly from replacing natural gas-driven compressors with electric-driven ones. The "Other Manufacturing" subsector contributes the next largest share, 450 Kt CO_2e, and some reductions occur in petroleum refining. Although both of these subsectors could take advantage of cogeneration, there is little incentive to do so because provincial electricity prices do not increase much under the policies.

The residential and commercial/institutional sectors contribute more than the electricity sector to emission reductions, though less than industry. The residential sector contributes less to reductions than the commercial/institutional sector – reductions are 60 Kt CO_2e under the sector target and 330 Kt CO_2e under the national target. Because of Manitoba's hydro-dominated generation, households do not experience sharp increases in electricity prices under the policies; however, natural gas prices do climb by about 60%. In the sector target simulation, households save approximately $26 annually but pay $326 a year under the national target (these amounts

Table 6.5

Emission reductions and costs, Manitoba

	Reference case	Sector target simulation	National target simulation
Emissions in 2010 (Mt CO_2e)	25.1	18.9	19.5
Energy consumed in 2010 (PJ)	357	302	300
GHG intensity in 2010 (t CO_2e/person)	19.3	14.6	15.0
Emissions relative to 1990[a] (%)	126	95	97
Emissions reduced in 2010 (Mt CO_2e)	n/a	6.2	5.6
Total cost (billion $)	n/a	2.8	1.9
GPD impact (%)	n/a	n/a	-1.4

a 1990 GHG emissions in Manitoba are 20 Mt CO_2e.

Figure 6.10

Contribution of sectors to emission reduction in Manitoba, 2010

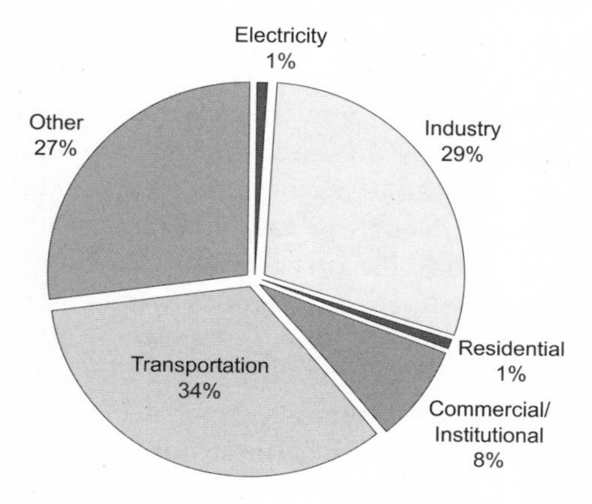

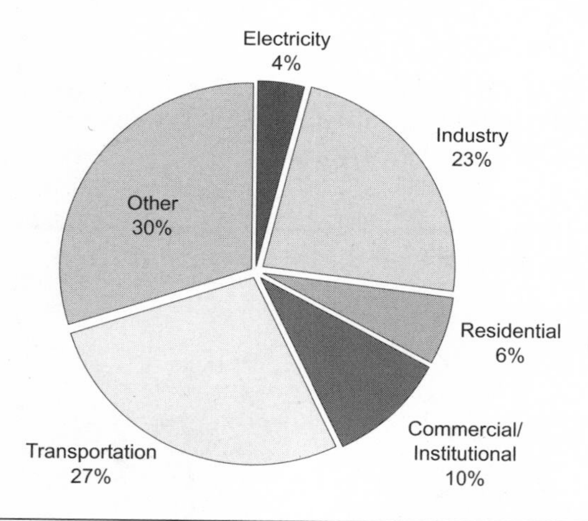

translate into a total saving of $80 million in the sector target simulation and costs of $1 billion in the national target simulation). The commercial/institutional sector sees only net benefits in reducing emissions. Total costs savings are $180 million under the sector target and $420 million under the national target, with reductions of 520 Kt CO_2e and 580 Mt CO_2e respectively.

GHG reductions in electricity are very small, between 100 and 200 Kt CO_2e, and occur from the small amount of fuel switching from natural gas and oil to hydro. Fewer reductions occur in the national target simulation.

Ontario

Ontario has both the largest population and the largest quantity of emissions of any region in Canada, 190 Mt CO_2e in 1996; however, it is expected to lose its status as the largest source of emissions to Alberta by 2010. At 17 Mt CO_2e per capita in 1996, its GHG intensity is lower than the Canadian average despite having a sizeable concentration of heavy industry. Its low GHG intensity for transportation, 4.6 Mt CO_2e per capita, is considerably less than that of the western provinces, where it exceeds 8 Mt CO_2e per capita. The greatest contributors to emissions in 1996 were industry, transportation, and electricity. By 2010 transportation is expected to exceed industry, even though growth in the transportation sector is relatively less compared with that of many regions. Industry emissions do not grow substantially.

Figure 6.11 shows the share of emissions by sector and their growth in the reference case. Emissions growth is greatest in electricity and transportation followed by "Other" (mainly nitrogen oxides and methane from agriculture). The commercial/institutional sector includes landfill gas emissions, which are quite large: Ontario has the majority of Canada's landfills (29 sites), which emit 3.5 Mt CO_2e a year. Transportation emissions grow strongly from population growth, though per capita GHG intensity does not grow as much in Ontario as in other provinces.

Electricity generation in Ontario is split between fossil fuel-based and GHG-free sources. However, the future trend is toward greater fossil fuel use. For safety reasons Ontario Hydro began a shutdown and maintenance

Figure 6.11

Reference case emissions by sector, Ontario

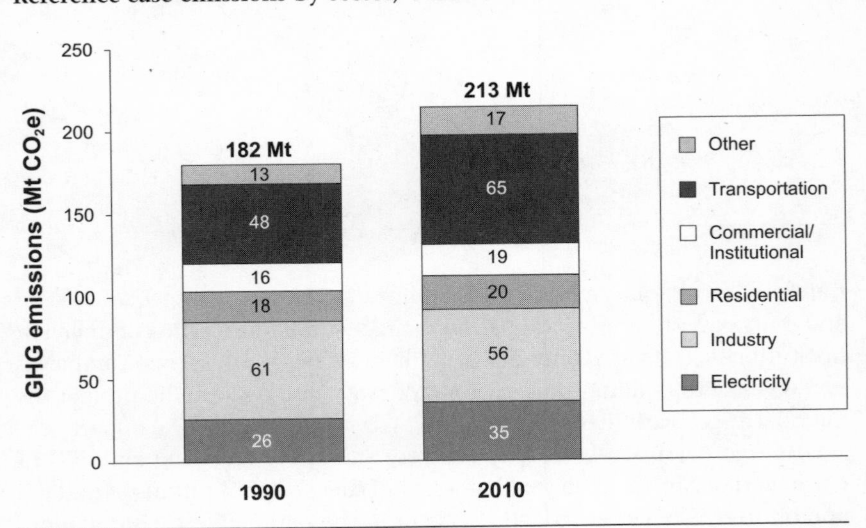

program of a number of its nuclear reactors in 1997. This program has resulted in more fossil fuel-based generation, mainly coal. Not all nuclear units are expected to return to production in the simulation period. Growth in demand for electricity is stronger than it is for Canada as a whole – 1.3% annually as opposed to 1.0%. Thus, electricity sector emissions grow by 36% between 1990 and 2010.

In contrast, industry emissions actually decline slightly over the same period. Many GHG-intensive industries – such as iron and steel, petroleum refining, "Other Manufacturing," and industrial minerals – do not expect strong growth in production. Also, actions by a few companies make significant impacts on reference case emissions. For example, the producer of adipic acid, used in nylon production, will eliminate its nitrogen oxide emissions and thus reduce 10 Mt CO_2e a year, almost 20% of total industrial emissions. Ontario firms that produce magnesium have also stated that they will eliminate sulphur hexafluorides (SF_6) from their processes before 2010. Because SF_6 has a very high CO_2e equivalency (1 kg SF_6 = 23,900 kg CO_2), the impact is also substantial.

GHG Policy Impacts

Given its share of the Canadian population and economic output, it is not surprising that Ontario contributes substantially to total emission reductions, at significant cost and GDP impact. As shown in Table 6.6, Ontario's emissions are reduced by between 17% and 20%, 48 Mt CO_2e in the national target simulation and 37 Mt CO_2e in the sector target. This percentage reduction is the lowest of all regions in Canada, but because of high total emissions Ontario contributes 25% to national emission reductions. The total cost for these reductions is $18-$23 billion, the higher value reflecting the sector target. Under the sector target, though, the industrial

Table 6.6

Emission reductions and costs, Ontario

	Reference case	Sector target simulation	National target simulation
Emissions in 2010 (Mt CO_2e)	213.1	176.6	165.0
Energy consumed in 2010 (PJ)	3,595	3,150	3,075
GHG intensity in 2010 (t CO_2e/person)	16.2	13.5	12.6
Emissions relative to 1990[a] (%)	117	97	91
Emissions reduced in 2010 (Mt CO_2e)	n/a	33.6	46.5
Total cost (billion $)	n/a	23.0	18.0
GPD impact (%)	n/a	n/a	-2.8

a 1990 GHG emissions in Ontario are 182 Mt CO_2e.

sector is forced to make costly reductions that can be abandoned under the national target.

The significant change in what the electricity sector contributes to total reductions can be seen in Figure 6.12; electricity changes from contributing no reductions in the sector target simulation to supplying 32% of the total in the national target simulation. In the sector target simulation, reductions from actions are offset by a 3% increase in electricity production, which

Figure 6.12

Contribution of sectors to emission reduction in Ontario, 2010

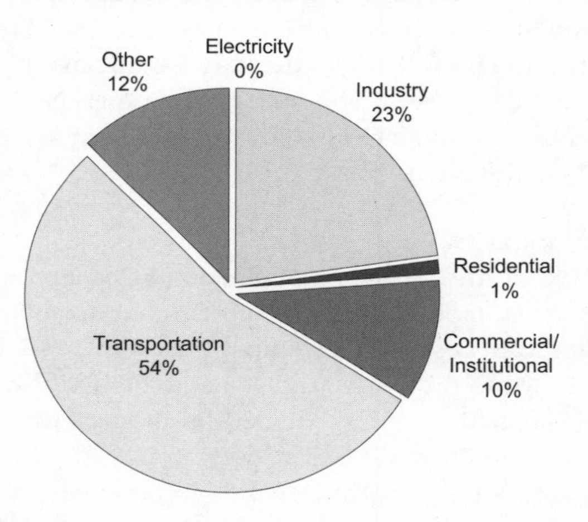

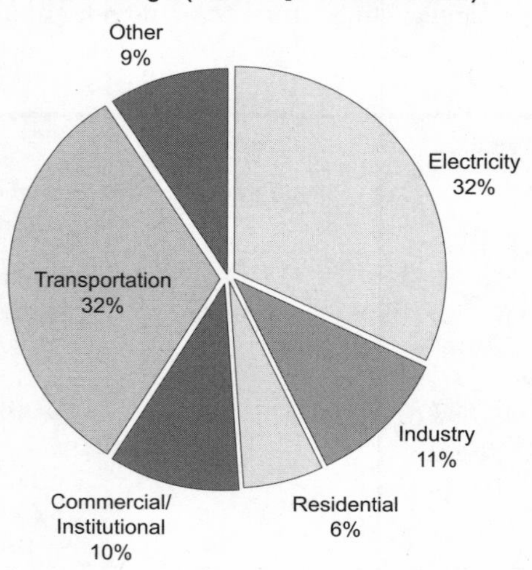

reflects a shift in electricity use from 21% to 25% of Ontario's total secondary energy consumption. In contrast, generation falls by 4% under the national target because the relative change in energy prices reduces demand. This reduced generation contributes to a decrease in emissions and even more emission reductions occur due to a shift from coal-fired generation to natural gas and biomass. Electricity price increases are relatively high under the national target (22%), though not nearly as high as those in Alberta and Saskatchewan.

As seen in Figure 6.12, the transportation sector makes significant reductions, particularly under the sector target, where it provides 54% of Ontario's total. As a province with large urban centres, Ontario has a high reduction potential: emissions decrease by 23% under the national target and 31% under the sector target. The largest share of these reductions occurs from transit improvements, particularly to pricing and infrastructure – 26% of the total. Speed-control measures contribute another 17% and efficiency standards 13%. All reductions in the sector cost between $7.2 billion and $11.7 billion, the higher cost occurring under the sector target. These costs are equivalent to $368 per household per year (sector target) and $226 per household per year (national target).

Even though the industrial sector is responsible for the greatest share of Ontario's emissions, reductions are more expensive relative to those in other sectors, putting this sector's contribution behind those of transportation and electricity. Reductions are 5.1 Mt CO_2e and 8.4 Mt CO_2e for the national and sector targets respectively and are made at a cost of $2.8 and $13 billion. Ontario's industrial reduction opportunities appear to be relatively high cost compared with those in other regions; many only occur between $120 and $300/t CO_2e reduced.

Industry reductions are also more diversified than in other regions. The majority, 2.1 Mt CO_2e, occur in the natural gas sector. As in Manitoba, all of Ontario's natural gas emissions are related to transmission and switching to electric-driven compressors and employing leak detection programs. The "Other Manufacturing" sector contributes the next largest share, 780 Kt CO_2e, largely from auxiliary equipment and steam efficiency improvements. No other subsector dominates the remainder of the reductions. The bulk of Canada's cement production occurs in this province, and, while significant efficiency improvements have already been made, further gains are possible at older, less efficient plants. Pulp and paper can switch to greater wood use for energy. Iron and steel makes some emission reductions, though as an energy-intensive industry, it has long pursued energy efficiency, and there is little room for fuel switching without structurally changing production. Of the remaining industries, some emission reductions occur in chemical manufacturing, mining, and metal smelting. Cogeneration use is already high in many industries, and little extra is adopted.

The residential and commercial/institutional sectors are fairly important to total reductions in Ontario because it is where much of the country's population and office space is concentrated. These sectors account for close to 16% of Ontario's emission reductions under the national target, a total of 4 Mt CO_2e. These reductions cost $1.7 billion and $6 billion for the commercial/institutional and residential sectors respectively. Key abatement actions are the capping and use of landfill gas, and improved efficiency in HVAC services and in building shells. Only negligible residential sector reductions are made under the sector target (490 Kt CO_2e). Under this policy simulation, the residential sector receives some cost savings ($680 million total or $21 per household per year).

Quebec

Even though Quebec has the second largest population among Canada's regions, its dependence on hydroelectric power puts it a distant third behind Alberta in terms of total GHG emissions. Between 1971 and 1992, electricity's share of Quebec's energy balance rose from 19% to 41%, while petroleum dropped from 74% to 42%.[9] The majority of homes and commercial buildings use electric heating, and industry uses more electricity than its counterparts in other provinces.[10] As a result, emissions in 1996 were 87 Mt CO_2e, and per capita emissions were 12 t CO_2e, the lowest in the country.

Figure 6.13 shows emissions by sector and their growth in the reference case. The industry, transportation, and commercial/institutional sectors emit the most GHGs. Per capita emissions in all sectors except for transportation and "Other" are expected to decrease over time. However, with population growth, only the residential and electricity emissions actually decline. The residential sector is expected to show a significant reduction in emissions: in addition to the efficiency improvements witnessed across Canada, electricity makes further inroads into energy use in the sector. Electricity sector emissions show a decline even from the small amount produced in 1990.

Quebec's industrial sector shares many similarities with Ontario's: the sector is diversified, and many energy-intensive industries do not expect significant growth in production. Growth in cement production declined from 1995 to 2000 and is expected to recover only marginally. As a result, energy use and consequent emissions are below 1990 levels in the reference case by more than 18%. Stronger growth is experienced in chemicals manufacturing and metal refining. The latter is quite important because it makes up about 40% of Quebec's industrial GHG emissions. Nevertheless, Quebec magnesium producers, like those in Ontario, will curtail sulphur hexafluoride emissions. This action will keep the growth in industrial emissions to only 12% between 1990 and 2010; without this, these emissions would grow by 34%.

Figure 6.13

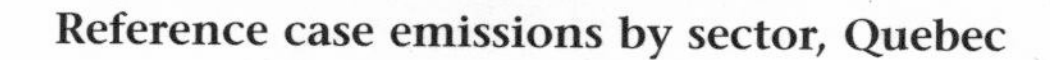

Reference case emissions by sector, Quebec

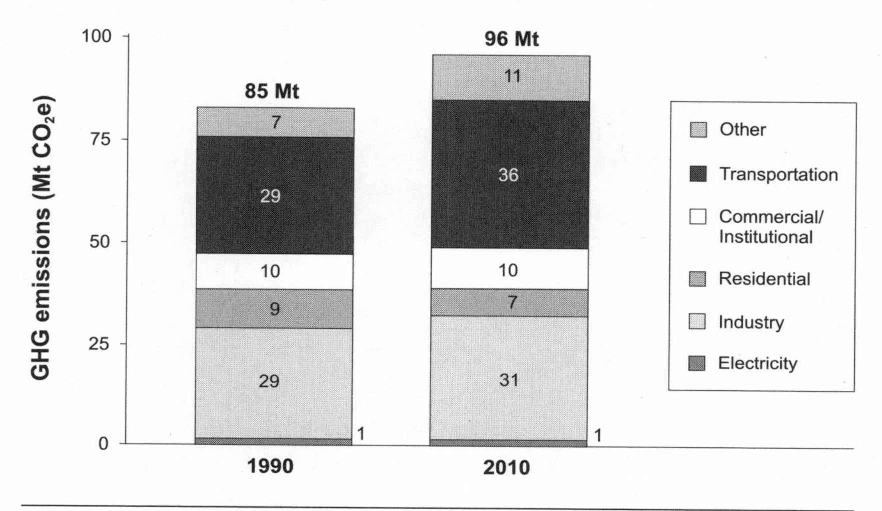

As in Ontario, per capita transportation emissions do not grow as strongly as they do in Western Canada. Emissions grow mainly from population growth, but decreasing fuel efficiency and increasing travel distances contribute as well. Cars are responsible for 70% of emissions, and trucks account for most of the remainder.

GHG Policy Impacts

For its population size, Quebec does not contribute significantly to national emission reductions: only 13%. Instead, actions are taken in other regions since they would be relatively expensive within Quebec – it has already turned to GHG-free energy to satisfy many of its energy needs. As shown in Table 6.7, its total direct costs are moderate, and compared with other provinces its GDP is less affected by reductions: the impact is half that of Ontario. Emissions in Quebec are reduced by approximately 20%, 19 Mt CO_2e in the national target simulation and 20 Mt CO_2e in the sector target simulation. Quebec's potential to reduce GHG emissions is greatest in the transportation sector, followed by the industrial sector (Figure 6.14). Hence, total reductions are slightly greater in the sector target, where transportation and industry are pushed for further emission reductions.

Even though Quebec's per capita GHG intensity in transportation is already the lowest in Canada, our results suggest that emissions can be reduced even further so that per capita intensity is lower in 2010 than in 1990. As in Ontario, a high level of urbanization in Quebec allows significant

Table 6.7

Emission reductions and costs, Quebec

	Reference case	Sector target simulation	National target simulation
Emissions in 2010 (Mt CO_2e)	96.2	76.2	77.6
Energy consumed in 2010 (PJ)	1,660	1,391	1,413
GHG intensity in 2010 (t CO_2e/person)	11.8	9.4	9.6
Emissions relative to 1990[a] (%)	113	90	91
Emissions reduced in 2010 (Mt CO_2e)	n/a	19.9	18.6
Total cost (billion $)	n/a	15.6	8.7
GPD impact (%)	n/a	n/a	-1.5

a 1990 GHG emissions in Quebec are 85 Mt CO_2e.

reductions from improving transit, 25% of the total. Speed-control measures contribute 17% and efficiency standards 13%. Transportation emissions are reduced by 23% and 30% at costs of $6.7 billion and $9.4 billion. The higher reduction and cost occurs in the sector target simulation. The total costs are equivalent to annual costs of $310 and $434 per household.

Industrial reduction potential is small and relatively expensive. Industry contributes reductions of 6.6 Mt CO_2e at a cost of $7 billion under the sector target and 5.2 Mt CO_2e at a cost of $2.8 billion under the national target. No single industry dominates these reductions. With few major natural gas transmission lines, emission reductions in the natural gas subsector are slight, only 120 Kt CO_2e in 2010. Instead, the largest source of industry reduction is "Other Manufacturing," which contributes 1.7 Mt CO_2e, followed by the pulp and paper subsector, which contributes 1.25 Mt CO_2e. Although these subsectors are already electricity oriented, some movement away from fossil fuel use occurs. The potential of actions to reduce emissions in chemical manufacturing and iron and steel production is relatively small. Iron and steel production is largely electrically based, and chemical manufacturing has few significant reduction possibilities besides cogeneration. However, this action is not attractive because electricity prices are low and do not increase much under the policy simulations.

As in Ontario, the residential and commercial/institutional sectors in Quebec offer fairly significant reductions given Quebec's population size and commercial floor space, 20% of the total. Under the national target, 2 Mt CO_2e are reduced in the residential sector at a cost of $290 million, and 1.7 Mt CO_2e are reduced in the commercial/institutional sector, with savings of $1.1 billion. Key actions include the capping and use of landfill gas as well as greater energy efficiency in commercial and residential buildings. In the national target simulation, electricity prices increase by only

Figure 6.14

Contribution of sectors to emission reduction in Quebec, 2010

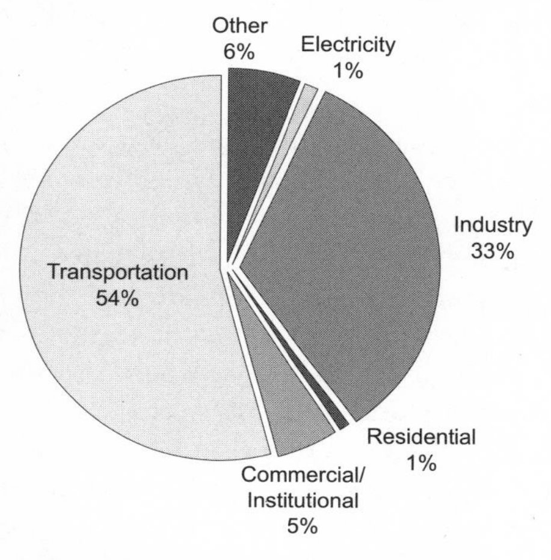

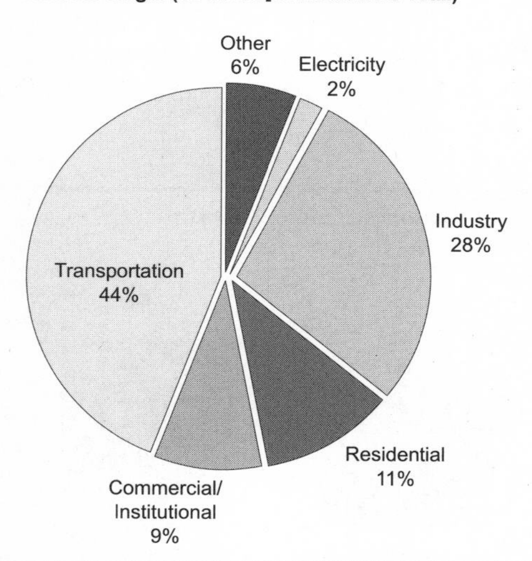

2%, while natural gas prices increase much more significantly, about 42%. As in all regions, reductions are negligible under the sector target (only 1.2 Mt CO_2e for both sectors) because these sectors are already close to reaching 6% below 1990 levels nationally. Under this policy simulation, the residential sector receives some cost savings, $290 million total or $9 per household per year. The commercial/institutional sector also realizes savings in this situation of $1.1 billion.

Few reductions are made in the electricity sector, but those that are made occur at a fairly high cost: reductions of only 200 Kt CO_2e and 300 Kt CO_2e are made under the national and sector targets respectively. Generation increases slightly in response to increased end-use demand encouraged by the policy, particularly under the sector target.

Atlantic Region

The Atlantic region in our study includes New Brunswick, Newfoundland, Nova Scotia, and Prince Edward Island, which were all modelled as a single region. This region produced 47 Mt CO_2e in 1996, about 7% of Canada's emissions, an amount that is on par with its share of the national population. The primary sources of GHG emissions are the electricity and transportation sectors, with industry a distant third. In the reference case, emissions grow primarily in the transportation sector; some growth also occurs in the electricity and industrial sectors. Compared with other regions, overall emissions growth is not strong over time – 14% from 1990 to 2010 (western provinces have emission increases in the 30%-50% range). Still, the growth in emissions is greater than that in population, which grows at 6%. Figure 6.15 shows emissions by sector and their growth in the reference case.

Although the Atlantic region has significant hydroelectric production, the bulk of this electricity is exported to Quebec and distributed from there.

Figure 6.15

Reference case emissions by sector, Atlantic region

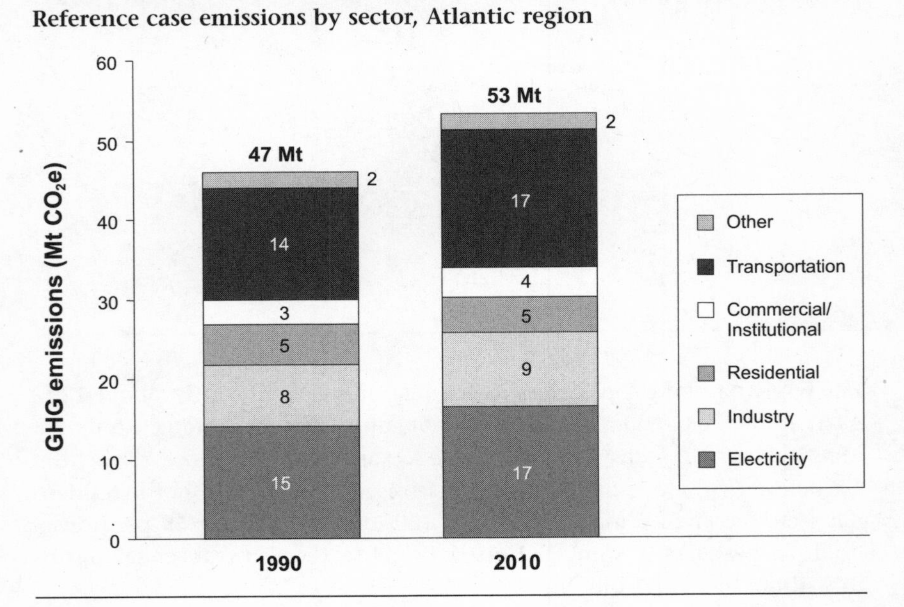

New Brunswick and Nova Scotia derive a lot of their electricity generation from fossil fuel sources: mainly coal and to a lesser extent oil. However, New Brunswick's use of nuclear energy mitigates the strong fossil fuel use in these provinces. Electricity GHG intensity in 1996 was 6.17 t CO_2e per capita, which lies between the intensities of hydroelectric-based provinces Manitoba and Quebec and the coal-dominated generation of Alberta and Saskatchewan. The reference case includes the construction of the Lower Churchill hydroelectric project, with an annual production of 17 tWh, starting in 2008. Nevertheless, emissions increase from the addition of fossil fuel plants to power the Voisey Bay nickel smelter and because New Brunswick is expected to replace some of its oil-fired plants with coal-fired plants. Some natural gas is assumed to be available for generation, though most of Nova Scotia's production is expected to be exported to US markets.

The largest industrial contributors to GHG emissions in this region are pulp and paper production and petroleum refining, which contribute about 20% each. The sizeable mining sector in Newfoundland and Labrador also contributes significantly, though mining is not particularly GHG intensive compared to other industries. However, if coal mining is also included, this source is significant, rising to 30% from 16%. Most coal-mining emissions are methane, which is significant given the region's largely underground shaft mining.[11] In general, industrial emissions are expected to grow due to increased economic activity from offshore oil and gas projects (such as Hibernia and Terra Nova) and Voisey Bay. This growth counteracts lower emissions from reduced coal-mining production. Fuel switching to natural gas is expected to significantly lower emissions only after 2010.

GHG Policy Impacts

Table 6.8 shows the resulting costs and reductions of the policy simulations. Emissions are cut significantly under the national target (31%), while under the sector target reductions of only 17% are made. Despite the significant decrease in the first case, economic effects in this region are small. At -0.9% the impact on GDP is the smallest in Canada. The total cost of reductions is $4.3 billion under the sector target and $3.1 billion under the national target. The region contributes almost 10% to national emission reductions. Reductions are greater under the national target because the Atlantic region offers low-cost abatement opportunities in the electricity sector, which are turned to (as in Saskatchewan and Alberta) when the sector requirements are relaxed. As shown in Figure 6.16, the relative contributions of the electricity and transportation sectors vary considerably according to the policy simulation.

Because CO_2 capture and sequestration was not considered to be an option in the Atlantic region in the 2010 time frame, the electricity supply sector cannot turn to this action to mitigate CO_2 from coal use. But there

Table 6.8

Emission reductions and costs, Atlantic region

	Reference case	Sector target simulation	National target simulation
Emissions in 2010 (Mt CO_2e)	53.1	44.2	36.8
Energy consumed in 2010 (PJ)	872	781	738
GHG intensity in 2010 (t CO_2e/person)	21.1	17.6	14.6
Emissions relative to 1990[a] (%)	114	95	79
Emissions reduced in 2010 (Mt CO_2e)	n/a	8.9	16.4
Total cost (billion $)	n/a	4.3	3.1
GPD impact (%)	n/a	n/a	-0.9

a 1990 GHG emissions in the Atlantic region are 47 Mt CO_2e.

are significant opportunities for emission reductions through fuel switching, and coal use declines considerably. Electricity prices increase by 15% in the national target simulation.

Transportation reductions occur mostly in passenger transportation. As a whole, emissions are 19% less than the reference case (under the national target), improving GHG intensity to 5.5 t CO_2e per person. Speed-control measures contribute about 20% to these reductions, while efficiency standards contribute 12%. Transit improvements provide only 9% of transportation reductions, compared with reductions of 25% in central Canada. Emission reduction costs are similar for both the national target and the sector target, about $4 billion (about $300 per household annually).

Emissions decline considerably in the industry sector under the policy simulations. A total of 2.5 Mt CO_2e is reduced at a cost of $1.1 billion under the national target, its largest contribution. Reduced coal production from lower coal use in electricity generation is responsible for 40% of these emission reductions. The cost described above takes into account lost revenue from lower production. The remaining emission reductions occur from actions in coal mining, metal and nonmetal mining, pulp and paper production, and "Other Manufacturing." The capture and use of coal-bed methane reduces 110 Kt CO_2e.[12] Mining actions contribute 480 Kt CO_2e, mainly by switching from heavy fuel oil to electricity. Fuel switching is also responsible for most pulp and paper and "Other Manufacturing" reductions. In the first case, switches are made over to wood waste as well as to natural gas (away from oil).

There are substantial emission reductions in the residential and commercial/institutional sectors: 34% and 19% under the national target for residential and commercial/institutional, respectively. Many reductions are due to fuel switching as natural gas becomes more readily available to the

Figure 6.16

Contribution of sectors to emission reduction in the Atlantic region, 2010

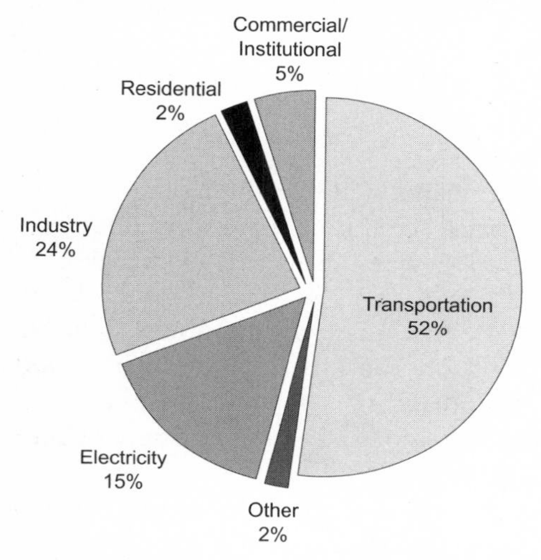

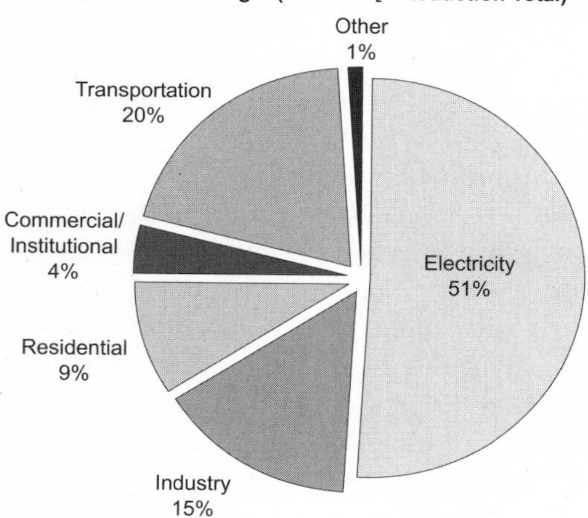

domestic market in the Atlantic region. Efficiency improvements also contribute. Nevertheless, these sectors contribute less to total emission reductions than other sectors – only about 13% combined – under the national target, its strongest impact. Under the permit-trading scheme of the national target, households would be faced with a 15% increase in electricity prices and a 42% increase in natural gas prices (across the whole Atlantic region). These increases are less than a corresponding increase in the price of oil, a more carbon-intensive fuel, which dominates residential and commercial/institutional fuel use. A strong switch away from oil is thus stimulated. Under the sector target, the residential sector saves $100 million ($17 yearly per household) and the commercial/institutional sector saves $20 million. Commercial/institutional reductions of 720 Kt CO_2e are made under the national target at a cost of $250 million.

Conclusion

What does this information tell us about our two opening questions: Does any one region bear a significant share of the burden of reducing GHG emissions? What can these results suggest about regional vulnerabilities and strengths?

In response to the first question, the results indicate that the costs of GHG emission reductions can vary significantly by region, in terms of both per capita costs and total costs. On a per capita basis, the Prairie provinces of Alberta and Saskatchewan, and to a lesser extent Manitoba, bear the greatest cost burden in the national target simulation. Both of the modelled policies bring out regional differences; however, they are accentuated in the national target simulation for a couple of reasons. First, the electricity sector contributes more to reductions. This contribution is significant because the GHG intensiveness of this sector varies greatly by region, and reductions are concentrated where they are most cost effective (in coal-dominated regions), so costs increase disproportionately for Alberta and Saskatchewan and are reduced for British Columbia and Quebec. Electricity price increases are also different by region and have considerable implications for the degree, type, and cost of actions in the other end-use sectors. This further differentiates costs and reductions between regions. Second, in contrast to electricity, patterns of emission reductions and per capita costs in transportation are fairly similar across regions (each region reduces transportation emissions by about 20%). Reducing the importance of these reductions results in more reductions made in dissimilar sectors.

The second introductory question has already been partly answered. A key source of cost variability is the dramatic difference in electricity supply between regions. Coal-intensive regions can make significant reductions at a fairly low cost. Thus, in order to reach the Kyoto-size target, Canada will have to rely on reduction opportunities that are concentrated in certain regions, particularly if it is to reach the target as cost effectively as possible.

The assumption thus far, however, is that the costs of meeting our Kyoto commitment are allocated to each region based on the size of its contribution, and this contribution is determined by a national cost-effectiveness analysis. There are many policy mechanisms for burden sharing and for initiating cost-effective action. These policies are presented in the next chapter.

7
Domestic Policy Options

Reducing GHG emissions is one of the major challenges of environmental sustainability facing humanity.[1] Although policy making in any domain is rarely easy, the objective of environmental sustainability has several attributes that make it a special concern for policy making in modern democracies.[2]

First, it is difficult for people to connect their actions as consumers with the local environmental impacts that concern them as citizens. Many of these local impacts are hidden from view. For example, waste generation and disposal, such as landfills and industrial production, are generally located outside urban centres. Also, many by-products of consumption, even the most potentially harmful, are unseen effluents and emissions with risks that relate to persistence and long-distance transport, effects that cannot be readily detected by experts, let alone the average citizen. GHGs fit into this category.

Second, the environmental effects of human activity are increasingly global and intergenerational. This extension makes it difficult for the current generation in one region to connect its pursuit of well-being and security with the potential repercussions for the well-being and security of future generations and even for current inhabitants of the other side of the planet. The highly uncertain nature of these distant effects diminishes the public and political will to take action.

Third, because humans have evolved an economic system that treats the environment as a free and unlimited waste receptacle, there will be substantial transitional costs in shifting that system toward lower flows of by-products such as GHG emissions. These costs must be incurred at a time when the options for government policy have narrowed. Governments cannot pay all of these costs from existing revenue. Regulations that raise costs are seen as heavy handed. Tax reforms designed to change behaviour in ways that will reduce environmental damage, even reforms that do not increase the net tax burden, are met with antagonism by the public and media in what is usually a polarized political climate. We seem to have entered an era in which governments are reluctant to set rules or adjust costs, opting instead to lead by example and form voluntary partnerships.[3] Some wonder how this restricted power can possibly effect the significant changes required to shift the economic system onto a more environmentally sustainable path.

Because of these special challenges of environmental policy making, our estimation of the costs of reducing GHG emissions would be incomplete if it did not also explore how, once these costs are recognized, Canadian society could begin to pursue the GHG path presented in this analysis. That is the purpose of this chapter.

In estimating the costs of reducing Canada's GHG emissions, we applied a simplifying assumption about the policy instrument that government would use to reduce these emissions. With the national target (Path 4), we tested an economy-wide emission cap and tradable permit system. Applied to the entire country, this policy can provide a single financial signal to all decision makers in the economy and should improve the chances of an economically efficient outcome. We also assumed that the policy causes no transfers of wealth, even though some sectors and regions face higher marginal costs than others.

The ideal policy, from an economic efficiency perspective, may be politically difficult to implement for any number of reasons. Some relate to general problems in policy making, such as public and media suspicions when governments intervene with economy-wide policies. Some relate to the special challenges of environmental policy making, notably the difficulty of getting the public to understand and support the costly or otherwise unpopular aspects of a policy that nonetheless is applied in pursuit of goals that have substantial public support. Some relate to the administrative difficulties of different types of policies; for example, a cap and permit-trading system may be complex to implement as an economy-wide instrument applying to all consumers and businesses. Some relate to the concentration of impacts on particular sectors and regions and even unintended transfers of wealth that result from the policy itself. Finally, some relate to the great uncertainties about the pace and character of technological developments and the potential for significant shifts in consumer preferences.

In this chapter we provide a more complete representation of the spectrum of policy options in order to compare other available options with the simplified policy assumption, which we made for the initial purpose of estimating costs. We then use this general representation, or policy framework, to comment on some of the current policy thrusts of governments in Canada and elsewhere and to forward our propositions for Canada's GHG policy emphasis.

A Framework for GHG Policy Options

Traditional Range of Policy Options

Figure 7.1 provides a framework for the broad categories of policy instruments for pursuing environmental improvement. These instruments are arranged along a continuum in terms of their *degree of compulsoriness.* We deliberately use this admittedly uncommon term because it best expresses the extent to which a certain behaviour is required by an external force.[4] A fully compulsory policy specifies exactly what must be done, and noncompliance is not an option because of the severity of the penalty. A less compulsory policy may require action by society in aggregate but confers some degree of flexibility for the firm or household. Policies are fully noncompulsory if the firm or household has the option to do nothing without suffering any negative consequences.[5]

Noncompulsory policies are on the left side of the continuum. They include policies in which government facilitates or initiates the development and dissemination of information (research and development, advertising, labelling, certifying, providing demonstration projects) that might influence the decisions of households, firms, and perhaps other levels of government. Government might provide information that convinces businesses or consumers of financial gain from actions that also improve the environment. Or government might provide information that convinces businesses and consumers to take actions for moral rather than financial reasons – the desire to contribute to a cleaner, environmentally sustainable world. Finally, government might lead by example, taking actions where it has direct control (land, buildings, vehicles, employees, publicly funded research and development, state-owned corporations) and hoping that others will follow or form partnerships with it for moral reasons and/or financial self-interest.[6]

Figure 7.1

Continuum of policy instruments according to degree of compulsoriness

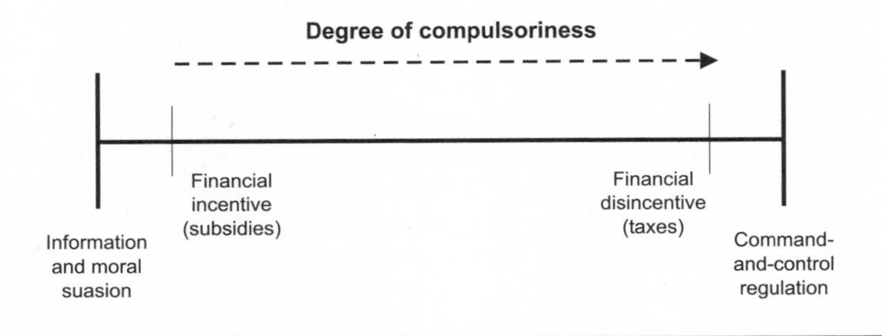

This left side of the continuum represents a decentralized approach to policy making and a noncoercive role for government. Government is seen as a facilitator, an encourager, and an information provider in order to convince members of society to act in their own domain. Thus, consumers might include environmental performance alongside financial considerations in product selection. This approach may be associated with green marketing, eco-labelling, and even environmentally motivated consumer boycotts. Unions might include environmental criteria in their negotiations with employers. Shareholders might include environmental objectives alongside other ethical criteria in their investment choices. Corporate managers might apply triple-bottom-line principles (balancing financial, environmental, and social objectives) in corporate decision making.

We also place financial incentives (subsidies) on this side of the continuum, although not at the far left. These are grants, loans, tax credits, and similar policy instruments that improve the financial returns for consumers, businesses, and even other levels of government that take actions to improve the environment. An example of the latter case is the application of environmental performance criteria for the awarding of government infrastructure grants for roads, transit, water, and other public undertakings. Given that financial incentives do not imply compulsory action – the business or consumer or local government can always opt not to participate – one could argue that they should be placed at the extreme left of the continuum. However, because they require funds that government must acquire from somewhere else in the economy, presumably in a compulsory manner, we place this policy instrument a bit to the right from the fully noncompulsory approaches.

Traditional command-and-control regulations are on the right side of the continuum. They include mandatory building codes, equipment specifications, vehicle and appliance standards, technology requirements, emission limits, prohibitions on the production and use of certain chemicals, and other kinds of regulations. These policies are enforced by financial penalties and even legal sanctions for noncompliance and are therefore at the extreme end of the policy continuum. Usually, the financial penalties are severe, and the regulatory application is specified (e.g., by plant or technology type or emission level) in a way that leaves virtually no decision-making flexibility.[7] For this reason regulations have often been criticized as an unnecessarily expensive way of achieving environmental objectives. The policy is insensitive to cost differences of pollution abatement between different consumers and different firms or between locations. Usually, all participants are required to behave in a similar manner (technology choice, emission level) even though some may be able to do more at lower costs than others. To economists this approach prevents application of the equimarginal principle, under which total costs of environmental improvement are minimized because every agent has the flexibility to reduce pollution only to the point where their incremented cost of pollution reduction is equal to the cost of other agents.

In spite of this potential for economic inefficiency, command-and-control regulations are still the dominant environmental policy tool of governments. This is because such regulations provide greater certainty for the regulator of the amount of emission or effluent or other form of environmental disturbance that will occur. This certainty is important when total emissions must not exceed local assimilative capacity or when dealing with highly toxic substances that must be completely eliminated.

Environmental taxes, such as GHG emission taxes, are unit charges for emissions that force firms and households to pay for some or all of the damages that they cause to the environment. As originally conceived, the charge per unit of emission should reflect the monetary value of the incremental damages caused by the emission, but in practice this may not be the case. Governments may decide to set the tax at a lower level for any number of reasons, one of them being to minimize adverse equity impacts.

Environmental taxes are situated toward the compulsory end of the policy continuum. True, they do not require one to take action to reduce environmental harm, and thus they decentralize decision making to consumers and producers. However, because either action or tax payment is required, there is a strong sense of compulsoriness. Indeed, the use of taxes as a policy instrument is often portrayed today as evidence of a coercive and intrusive government. Additional suspicion is generated by the common concern that taxes reflect government revenue needs rather than legitimate public policy goals.

As the above discussion suggests, progression along the continuum from left to right is associated not only with increasing compulsoriness but also generally with decreasing policy acceptability. Firms and households want to retain as much freedom of action as possible. A regulation may reduce that freedom significantly, with perhaps dramatic cost consequences. Even an environmental tax can have substantial cost consequences. Voluntary programs are the most acceptable. Of course, households may be willing to see regulations applied to industry, and perhaps vice versa, but here we are referring to the willingness of either party to be subject itself to taxes or regulations.

Governments recognize this, or soon learn from painful experiences, and naturally prefer the left side of the continuum. Unfortunately, the continuum also appears to correlate roughly with another dimension – policy effectiveness. Compulsory policies tend to be more effective even if unpopular. Voluntary programs and information dissemination are not unpopular but may be less effective than advocates sometimes claim.

The challenge for policy making is to find policy designs that do better than others in trading off the conflicting factors of the continuum. The ideal policy, then, is highly effective yet passes the test of public and corporate acceptability (i.e., is not seen as unfair or overly compulsory). Some policies will do better than others in meeting this challenge.

Promising New Policy Instruments

Two new types of policy instruments that have emerged over the past decade are positioned in about the middle of the continuum of Figure 7.1 in that both include some degree of compulsoriness yet allow considerable flexibility to firms and households. We refer to one as environmental tax shifting and the other as market-oriented regulation.

Environmental tax shifting involves levying environmental taxes and recycling all of the tax revenue as rebates to those who pay the taxes or as reductions in general taxes or charges. While the idea of governments linking the increase of some taxes with the decrease of others is not new, its application to environmental policy has only recently been articulated as a comprehensive strategy. Some environmental advocates are supportive of environmental tax shifting as a key component of environmental policy.[8] They hope for greater public acceptability by linking the increase in environmental taxes with the reduction in unpopular taxes. They also hope that transparency in the flow of revenues will mitigate the usual public suspicions about inefficient government use of tax revenues. Thus far there have been several tentative initiatives in environmental tax shifting in most industrialized countries, as well as more serious applications by some European governments.[9] Modest initiatives include policies such as deposit-refund schemes and *vehicle feebates,* while more ambitious initiatives under consideration involve the application of greenhouse gas tax revenue to reduce government payroll charges, income taxes, or other broad levies.[10]

The challenge of finding a balance between acceptability and effectiveness, combined with a concern for economic efficiency, has led some to advocate the assignment of private property rights to common property resources (which are frequently the focus of environmental damage). In theory this assignment of property rights would enable parties to use the legal system to determine the appropriate compensation for emission damages and even set the appropriate level of emissions.[11] In practice there are substantial challenges with this approach. First, the initial assignment of property rights has implications for the distribution of wealth and thus for an efficient outcome. The party who is initially awarded the property right to the common property resource experiences an increase in real income that may affect the marginal price at which the party is willing to trade that property right. Second, some parties (such as a widespread coalition of common interests) may face high transaction costs for use of the legal system, and these costs will also affect an efficient outcome.

Because of these and other challenges, the assignment of property rights has been rarely applied as a means of allowing individual members of society to determine, via trading of property rights and financial exchanges, the appropriate total level of emissions. However, the idea of property rights can also be used in a less ambitious manner. That is, once society establishes – via some other mechanism of public choice – the desired target level of emissions, the rights to fixed levels of emissions can be allocated and traded to achieve the target as efficiently as possible. We refer to this as a market-oriented regulation.

Market-oriented regulation is a form of regulation in that an aggregate target, such as an economy-wide emissions cap or a level of technology market penetration, is compulsory. Also, all firms and households are involved in some way, and noncompliance has dramatic cost and/or legal consequences. However, market-oriented regulation is unlike traditional command-and-control regulation, and more like an environmental tax, in that participation occurs at the discretion of firms and households. Some may contribute to the achievement of the aggregate target by reducing emissions or acquiring the designated technology, while some may pay others to do more in order to make up for their unwillingness to reduce emissions or to acquire a technology.[12]

The best-known example of a market-oriented regulation is the *cap and tradable permit*. It sets a total emission limit, or cap, for whatever entity is being regulated – several firms, an entire country, or the globe. Shares of this emission limit are allocated as permits by some method (historical levels or grandfathering, auction, some combination of grandfathering and auction) to individual participants (businesses, provinces, even countries depending on the scope of the program). The regulation would include rules for decreasing the total cap over time and for determining how new participants in the program would access permits. The shares are tradable, in effect providing a specified right to pollute that can be traded just like any property right.

The cap and trade regulation has attractive features from a policy design perspective. First, as a form of regulation, albeit aggregate regulation, the policy has a high likelihood of achieving the environmental target. In contrast, voluntary programs and environmental taxes are substantially more uncertain in terms of target achievement because a specific outcome is not compulsory. The outcome depends on the shifting motives of firms and households (voluntary) or their responsiveness to changing prices (price elasticity in response to fiscal policies). Second, by allowing trading among participants, the cap and trade regulation can function like a tax in providing a uniform cost signal to all participants – the permit-trading price – that helps to minimize the total costs of achieving the target.[13] Third, as long as there is a positive price for permits, there remains an incentive for further innovations that profitably reduce emissions; in contrast, once one has satisfied a regulation that specifies a certain technology or emission level for each emitter, there is no incentive to do more. Fourth, the policy has the flexibility of a tax in allowing each emitter to determine the optimal combination of emission reduction and permit purchasing. In this sense it does not appear to be as compulsory as conventional command-and-control regulations. Thus, the policy is situated, along with environmental taxes, in about the middle of the policy continuum of Figure 7.1.

The most noteworthy application of a cap and tradable permit regulation started with the amendments to the US Clean Air Act in 1990. Sulphur emissions from specified electricity generation plants in the United States were subject to a cap and tradable permit regulation with a first phase from 1995 to 2000 and a second, more ambitious, phase after 2000. This program has already resulted in substantial reductions in emissions, and the total cost of reduction has been significantly lower than anticipated.[14]

The US sulphur permit-trading policy has inspired a growing number of similar policy experiments in the United States and elsewhere. These experiments are usually focused on the control of emissions; thus, it is the emission itself that is capped, permitted, and traded. However, this approach can be generalized beyond emissions to regulations that specify some other attribute, such as the type of technology or the form of energy that is used. Two noteworthy policy innovations of this type are the renewable portfolio standard in electricity generation and the vehicle emission standard in the automobile sector.

The *renewable portfolio standard* (RPS) is a form of market-oriented regulation that requires providers of electricity to guarantee that a minimum percentage of their electricity is produced using renewable energy. Emerging in the 1990s, this policy does not specify the production choice of each producer, only the aggregate market outcome. Each producer decides if it will generate some renewable electricity itself or purchase credits from a renewable electricity generator. If markets function effectively, then producers trade among themselves (as with the tradable emission permits), and only those with the lowest costs generate renewable electricity for the entire market. There is no guaranteed price for renewable electricity, only a guaranteed market share. Not having a guaranteed price sustains the competitive pressure for cost reductions because any reduction in the cost of producing renewable electricity will lead to higher returns and/or larger market share for the individual renewable producer. Because each purchaser of electricity is paying a blended price, comprised of the new renewables along with the dominant conventional supply, the RPS has a negligible effect on current rates. RPS targets are modest initially, giving time for the market to adjust and for competitive pressures and commercialization to drive down the costs of renewables.

Governments have traditionally supported renewables with subsidies for research and development, supply price subsidies, tax credits, and information and voluntary programs. If desired they can be retained alongside the RPS. A recent survey of the RPS indicates that the policy has been formally adopted by nine states in the United States, three European countries, and Australia.[15] It has also been under consideration in legislation before the European Union and the federal US government.

The *vehicle emission standard* (VES) is another form of market-oriented regulation that requires automobile manufacturers to guarantee that a minimum percentage of vehicle sales meets different categories of maximum emission levels. The policy originated in California and is the central focus of that state's efforts to improve local air quality. Until the 1990s manufacturers claimed that extremely low- or zero-emission vehicles would be too expensive to build and, in any case, would not satisfy consumer demands for vehicle attributes such as acceleration, horsepower, safety, and fuel range. Then in 1990 the California Air Resources Board established its VES.[16] The policy requires that a minimum percentage of vehicle sales be in each low- or zero-emission category. Individual manufacturers of vehicles are expected to meet the standard as a fleet average of retail sales in California, but a flexibility provision allows for trade between manufacturers so that the total California fleet meets the standard even if individual manufacturers fall short. The policy gave manufacturers considerable lead-time between target setting (1990) and firm target dates (1998, 2003).

The VES has not been implemented without challenges, especially in the negotiations and debates surrounding the zero-emission vehicle category.[17] However, the California VES seems to have played a significant role in the recent emergence of revolutionary new vehicle technologies, notably electric-gasoline hybrids, battery-electric, and fuel cell-electric. Manufacturers now seem to be intent on outcompeting each other to capture this new market, which is reflected in recent research funding, commercialization efforts, and marketing strategies. The California legislation has been copied by New York, Massachusetts, Vermont, and Maine.

These two policies, the RPS and the VES, are of special interest for several reasons. First, by providing an adequate time frame along with firm deadlines and penalties, both policies seek to mobilize producers to make the long-term research and development effort needed to fundamentally change technologies. Second, both policies act at the nexus of new product development and mass commercialization. Research on technology diffusion indicates that technology costs can experience a dramatic decline once the scale of production surpasses critical thresholds, which is more likely with a guaranteed market share.[18] Third, both policies work to reduce costs by allowing producers to trade among themselves in achieving the aggregate, regulated outcome. Fourth, both policies provide an incentive for producers to rethink their marketing strategies; if producers can convince consumers to pay a premium for the value that they believe they receive from renewable electricity or low-emission vehicles, then the financial benefits to producers increase. The result is to mobilize producers to market "greenness" to consumers. Fifth, both policies can be directly linked to the environmental target. Denmark and the Netherlands present the RPS as one of the key policy components for the attainment of their GHG emission reduction targets. California has linked its VES to its local air quality objectives, but is currently passing legislation to include GHG objectives.[19] Sixth, both policies (like the sulphur cap and trade policy in the United States) apply respectively to just one sector of the economy, reducing negotiation challenges and perhaps increasing the chances of acceptability. At the same time, applying a different policy instrument and target to each sector increases the risks of transgressing the equi-marginal principle – undertaking high-cost emissions reductions in some sectors while ignoring lower-cost options in others. However, a pre-implementation assessment of the policy, combined with post-implementation monitoring and adjustment, can reduce this effect.

Assessment of Policy Options for GHG Emission Abatement in Canada

Environmental policy is almost inevitably a package of different policy instruments working in combination. Government policies reflect the competing preferences of politicians and interest groups, and these preferences are not static over time. Also, there is usually a life cycle to policy issues. In the early stages of issue awareness, governments are less likely to operate on the compulsory side of the continuum.[20] There is still too much uncertainty about the environmental risks and the public's perception of the issue for government to push ahead with policies that will impose costs on some.

GHG emission abatement policy is influenced by these factors. First, the GHG issue is emerging at a time when the public acceptance of government intervention is low relative to previous decades. Governments behave and talk today as if their authority to cause economic pain for environmental objectives is severely limited. Second, the GHG issue is still relatively new for the public and policy makers. Governments do not yet sense the groundswell of concern that would mobilize them toward the more forceful and effective end of the policy continuum.

While regulatory approaches have been downplayed, they have not been entirely absent. Ongoing processes since the 1970s to tighten various regulations have continued over the past decade. Governments now link these processes to their GHG actions. Examples include stricter energy efficiency standards for buildings, appliances, industrial equipment, and vehicles. But these regulations are not pursued as leading instruments for GHG abatement; their role is more to consolidate technology transformation by ushering out the least energy-efficient equipment, processes, and buildings.

The present period is dominated by governments seeming to take action while avoiding initiatives that would incur the kinds of costs for firms and households that appear to be necessary according to our cost estimates in this study. This preference favours information and voluntary policies over others.

How Far Can We Get by Emphasizing Information and Voluntarism?

The major focus of Canadian GHG policy over the past decade has been to explore how far the country might get with actions that are driven mostly by policies from the noncompulsory side of the policy continuum. At the federal level, the National Action Program on Climate Change (1993) and the subsequent First National Climate Change Business Plan (2000) are dominated by policies that seek to motivate actions by firms and households on the basis of self-interest or moral values. There are information programs (advertising, Internet resources, labelling) to inform businesses and consumers about the economic benefits of actions that reduce energy use or waste streams or other resource use and save money at the same time. There are calls for voluntary actions by firms and households that cater to moral satisfaction. There are educational programs for the public and schools. The government also offers itself as a facilitator for those willing to initiate action. This kind of support appears in many guises, from coordinating a voluntary emissions-trading program among industries to providing training programs for agriculture and forestry management. At the provincial and municipal levels, the approach is similar, although it varies somewhat from place to place. Provincial governments facilitate voluntary programs such as carpooling and efficient building design. At the municipal level, urban air quality and general urban livability (congestion management, land use planning) have long been pursued, but there has generally been little specific focus on GHG emissions, with the exception of information programs. This is changing with the growing efforts of the Federation of Canadian Municipalities Partners for Climate Change Program, as well as increasing Canadian municipal membership in the International Council for Local Environmental Initiatives. Now more than 60 municipalities have commitments to develop climate change action plans as part of their involvement with these two programs.[21]

A key federal government initiative started in the early 1990s is the Voluntary Challenge and Registry, in which industries are asked to reduce GHGs on their own initiative. The federal government also created the Office of

Energy Efficiency, which plays a lead role in providing information on energy efficiency benefits and subsidizes some efficiency programs. In addition, federal, provincial, and some municipal governments are focusing on those GHG emissions that they are directly responsible for by investing in efficiency improvements to public buildings, changing how they purchase vehicles and electricity, and launching similar initiatives.

Governments today appear to be more enthusiastic than ever about the emission abatement that can be realized simply by instilling greater economic efficiency awareness in firms and households where it is coincidental with environmental improvement. Using the general term *eco-efficiency* – in concert with new concepts for business such as *triple bottom line* (financial returns, environmental improvement, social well-being) and *the natural step* – advocates in industry, government, and environmental organizations suggest that a great deal of environmental improvement, including substantial GHG emissions abatement, can be attained with this ethic of combining voluntarism with profit maximizing.[22] Indeed, individual firms and industry associations have many anecdotal studies to show how they have taken actions that simultaneously improved economic performance and reduced GHG emissions. For example, the natural gas industry is working on dramatically decreasing the fugitive emissions of GHGs from leaks during the production, processing, and transport of natural gas.[23]

But are voluntarism and profit potential enough? Our cost analysis in the previous chapters suggests that, while meeting a target such as the Kyoto Protocol may not be cataclysmic to the Canadian economy, it will have significant financial and intangible costs to firms and households. These costs will be over and above the benefits that might be realized from eco-efficiency initiatives. Our results, and those of other models that incorporate consumer preferences, suggest that much more will be required. The national process in Canada also seems to recognize this by focusing more and more research on the middle and compulsory portions of the policy continuum, exploring in particular the effects of different designs for cap and tradable permit programs.

Another problem with voluntarism is the difficulty of verifying the contribution of such an approach.[24] Many claims are made. Indeed, it is in the interests of industry to list actions that it would have done anyway and perhaps convince government that more costly policies are not required. But if these actions were already in the reference case scenario (i.e., they would have occurred anyway), and that scenario shows Canadian emissions continuing to rise, then they do not move the country toward the lower-emissions trajectory that would meet the Kyoto or some subsequent target.

This is why some people claim that the best way to measure the effect of the voluntary program strategy is simply to see what is happening to emissions in aggregate. In this respect the federal government's *Canadian Emissions Outlook: An Update* (*CEOU*) of late 1999 showed that emissions continued to rise in Canada in the 1990s at a dramatic rate in spite of the government's Rio commitment and its largely voluntary National Action Program on Climate Change (NAPCC) in 1993. A hindsight analysis by the Pembina Institute suggested that this program had been ineffective and that companies had frequently listed reference case activity as emission reduction actions.[25] Nonetheless, *CEOU* claims that the NAPCC reduced emissions in 2000 by 35 Mt CO_2e from what they otherwise would have been. Verifiable empirical evidence has not been provided to back up this claim.[26]

Another concern is that, without the persistent financial signal of a GHG tax or a GHG cap and tradable permit system, emission reductions resulting from voluntary eco-efficiency can be quickly offset by economic growth and even a shift toward the products that have realized a gain in energy productivity because of the resulting lower energy costs that follow. There is considerable debate among analysts about the potential magnitude of this *rebound effect*.[27]

One hope is that a voluntary policy focus of government can accelerate society along an inevitable trend toward a much less energy- and material-intensive economy, the so-called *knowledge-based economy*. Dramatic environmental gains may follow as the focus of research and development shifts increasingly toward products and services that require less energy and materials. Again, however, there is much debate among analysts.[28] A related argument is that strong environmental regulations, as opposed to noncompulsory policies, give an economy a competitive advantage in the future world of ever more stringent constraints on pollution.[29]

What if Subsidies Are Added to Voluntarism?

In the energy crisis days of the 1970s and the electricity efficiency push of the 1980s, governments and electric utilities learned that information programs alone may have little influence on firm and household energy use decisions. Initially, governments and utilities funded audit programs in which energy experts conducted on-site analyses to identify profitable efficiency actions for homes and factories. But hindsight reviews revealed that only a few of these investments were ever undertaken.[30] More successful were programs that combined information such as audits with financial support. Governments have subsidized home insulation retrofits. Electric utilities have subsidized the purchase of efficient appliances, lightbulbs, and industrial motors.

However, funding is problematic. Electric utilities provided subsidies from their general revenues, but the legitimacy of this approach was eventually undermined as advocates for nonparticipants convinced utility regulators of the inequities of taxing all electricity consumers in order to transfer wealth to the participants of efficiency programs. At the same time, with ballooning government deficits in the 1980s and 1990s, public subsidies for energy efficiency actions were eliminated or cut dramatically.

As public finances have improved in recent years, many governments have once again been open to providing more public support.[31] While some of this support may be in the form of direct subsidies, especially for research and development and for infrastructure such as public transit, the government preference is to provide low-interest loans and tax credits for research, development, and commercialization of energy efficiency, fuel switching, renewable energy supply, and even CO_2 capture and sequestration. Tax credits have the advantage of being a less visible form of public subsidy since their effect is to reduce government tax revenues rather than provide direct financial transfers. Government infrastructure grants increase and decrease depending on budgetary and political cycles. Environmental performance can be explicitly linked to these grants; indeed, lower levels of government can be required to compete for such funds based on relative environmental performance as a way of *environmental leveraging* of infrastructure support. Another kind of leveraging is possible with revolving public funds that provide low-interest loans in support of building retrofits that improve energy efficiency, as the City of Toronto has demonstrated with its Better Building Partnership. This program and other GHG reduction programs are supported by the Toronto Atmospheric Fund.

Although electric utilities have decreased energy efficiency subsidies to businesses and consumers, they have simultaneously recognized the benefit of shifting their efforts to the manufacturers of energy-using equipment.[32] The lesson is that, once manufacturers have been induced to produce efficient products, they quickly become much more effective than government in informing consumers and influencing their preferences by virtue of the compelling need to ensure adequate sales of these new product lines.

Our estimate of the cost to Canadians of achieving a Kyoto target for GHG emission reduction is based on simulating in our model a financial signal that we refer to as the GHG permit price in a domestic cap and tradable permit system. This signal might equally be used to indicate the sizes of subsidies that firms and households require to compensate them for the costs of taking actions to reduce their GHG emissions. In theory, government could provide all of this money in the form of subsidies, collecting the necessary funds from general tax revenue. This was essentially the approach in the heyday of electric utility efficiency programs as utilities taxed all customers through regulated tariffs in order to pay subsidies to those who would undertake electricity efficiency investments. While governments in Canada (federal, provincial, and municipal) are showing an increased willingness to provide GHG emission reduction subsidies, they will be far less than the $45 billion required according to our cost analysis.

Governments are increasingly recognizing that they need to look at policies toward the compulsory end of the continuum. Environmental taxes and various forms of regulations – especially market-oriented regulations – are being examined from a number of perspectives: economic efficiency, incidence of impacts, and political acceptability.

What about Environmental Tax Shifting?

Environmental taxes, in particular the strategic application of environmental tax shifting, offer a way for governments to shift along the policy continuum as needed to achieve a target such as the Kyoto Protocol. However, environmental tax shifting must overcome the perceived risks to politicians of any tax adjustment other than a clear and simple tax decrease. Recognizing this risk, supporters of environmental taxes emphasize the reductions in unpopular taxes – such as income taxes and payroll charges – that environmental tax shifting enables.

But this is still a difficult sell. One reason is that those who might benefit from a tax shift are generally less aware and less mobilized to support such a change prior to receiving the resulting benefits. New companies that might emerge to take advantage of the tax shift are in their infancy. In contrast, those who would bear the cost of environmental tax shifting are more aware and mobilized. For example, industries with high emissions are well established, with communities and regions dependent on their survival. There is also considerable suspicion among the public and media that governments tinker with the tax system only to raise revenue, in spite of claims of revenue neutrality. Even though environmental tax shifting may be revenue neutral, thus providing net benefits for many, politicians are aware of how difficult it is to convince the public and are reluctant to take significant initiatives in this direction.

Given that environmental tax shifting is advocated by many economists and environmentalists, it may yet emerge in the mix of GHG policies. However, it may initially play more of a consolidating rather than a driving role, meaning that only modest taxes would be experimented with in support of other leading-edge policies and actions.

What about Economy-Wide Cap and Tradable Permit Systems?

Our simulation of the national target (Path 4) is an application of an economy-wide, domestic emission trading (DET) system. Depending on the permit allocation system (auction, grandfathering, hybrid), the incidence of DET (who pays more) can differ greatly. Different approaches to DET can also have different administrative costs. The DET mechanisms were tracked through a macroeconomic model in order to estimate some of these indirect effects, but we assumed that all money needed to purchase a permit was returned to firms and households roughly in proportion to their initial permit payments. Thus, our study showed less regional and sectoral impact than could occur under different rules of permit allocation. If all permits were auctioned, for example, sectors and regions that are currently high GHG emitters would experience substantial revenue losses and, with the resulting higher costs of production, significant indirect, macroeconomic effects. Instead, if permits were allocated according to initial emission levels (grandfathered), then sectors and regions that are currently high GHG emitters could benefit if their marginal costs of GHG emission reduction are lower than the national average.

We believe that, if Canada eventually implements a DET system, it will be one that minimizes to the greatest possible extent these possible second-order effects. We do not believe that the Canadian federal system, and current political realities in Canada, will allow significant deviations from this policy design criterion. In all likelihood this means that the DET system would favour grandfathered permits, but might require auction purchase of some percentage (say 25%) to ensure that grandfathered permits do not lead to windfall profits for those with the lowest marginal costs of GHG emission reduction. If we are correct, then the micro- and macroeconomic impacts that we report would be a good approximation of not just the direct costs but also the indirect regional and sectoral costs of a DET system in Canada.

Because a DET system provides a single financial signal and does not associate government GHG abatement policy with taxes, there is considerable government interest in this policy instrument, especially when focusing on a specific emissions target, as in the case of the Kyoto agreement. However, a DET system would operate very similarly to a tax in terms of its effect on energy prices. The energy price estimates presented in Chapter 4 may differ slightly between GHG tax and DET, but they should be in this range. Thus, applying a DET system to achieve the Kyoto target could increase gasoline prices by 40¢/litre and some residential electricity rates by 50% or more. These substantial price increases would likely evoke a negative public response. It is in anticipation of this response that some analysts are increasingly looking for policies that work in similar ways to DET but with a less direct and hopefully smaller use of price change to drive the needed GHG emission reduction actions. In this light some of the other market-oriented regulations may be of interest.

What about Sector-Specific, Market-Oriented Regulations?

The principles of a cap and tradable permit system can be applied at the sectoral level for setting targets for emissions, energy forms, or technologies. An emissions level for all sectoral participants might be set, with trading allowed after the initial allocation. Or a minimum market share for less polluting energy forms or equipment might be set, again with trading allowed for attaining the sector-wide minimum. While this sectoral approach would forgo the economic efficiency benefits of a unique, economy-wide financial signal, it may have a better chance of success by virtue of its less ambitious scope.

In fact, even the most frequently cited example of a cap and tradable permit system is an application to only one sector of the economy: the US sulphur emission policy for electricity generators. From a policy maker's perspective, it may be easier to establish a GHG cap and tradable permit system for the Canadian electricity sector alone than to apply such a system to the entire economy or all of Canadian industry. Such a sector-specific policy is currently under consideration for the US electricity industry and was initially supported by President Bush during his election campaign.

Other sector-specific, market-oriented policies, such as the RPS and the VES, are still relatively new, making it precarious to speculate on their broader potential for GHG abatement policy.[33] Nonetheless, it is difficult to ignore the possibilities. Examples of candidate applications would be to require manufacturers to achieve minimum market shares for high-efficiency lightbulbs, appliances, and certain industrial equipment such as motors. There are many possibilities, although unique issues of institutional context and market structure must be addressed in all cases.

Our Preferred GHG Abatement Policies

This review of policy options and the results of our cost estimation lead us to develop our key criteria for designing GHG abatement policy in Canada. We present these criteria and then follow with a suggested package of policies that, in our view, would best meet them.

Criteria for Policy Design

While our focus is on domestic policies, we assume that Canadian policy making is constrained at a minimum by the intensity of any US GHG policy commitment. This does not mean that Canada cannot act until the United States has made a national commitment to ratify the Kyoto or some subsequent agreement. Indeed, our preferred policies tend to be ones that governments are already applying at national and state levels in the United States, though not necessarily as part of an explicit GHG strategy. While some Canadian policies can only be implemented in an ambitious fashion if the United States and other key countries are also taking significant action, they can be implemented to some degree. Public leadership in Canada may be judged by the ability of our governments, at different levels, to recognize that they can develop and implement innovative and forward-looking GHG abatement policies even in the absence of a national US commitment.

The GHG challenge is long term and as such calls for policies that operate on the long-term determinants of GHG emissions. In this sense the Kyoto target could have been a useful motivator if society were ready to start in 1997 to move on a target of 2010. However, to start a serious effort just half a decade before this or any deadline could lead to an exaggerated effort that fosters short-term, unenduring preference changes and readily available but costly technology changes such as efforts to accelerate the turnover of relatively short-lived equipment. Policies with a long-term focus would instead seek to influence the fundamental character of technological innovation, the forest and agricultural land base, and the gradual evolution of urban form and infrastructure with policies such as research and development support (for efficiency and cleaner fuels), warnings about future higher financial penalties for GHG emissions, and a shifting emphasis in urban planning and infrastructure funding. Some of these policy instruments may act too slowly to realize the Kyoto target on time. However, their impacts may be that much greater and at lower costs in the period after 2010. In deciding whether or not to delay meeting a target, one must assess the risks of incurring high costs in the short term only to see cheaper options emerge as time passes. At the same time, complete inaction in the present is unlikely to create the incentives and other signals necessary to motivate the development of cheaper options in the future.

GHG policy should reflect the information from cost analyses, such as ours, for determining how much can be realistically achieved in what time frame by firms and households. First, this requires an understanding of how the sectors of the economy are integrated. Anecdotal studies of profitable eco-efficiency initiatives can be misleading if they do not account for simultaneous changes in other sectors.[34] Second, while information campaigns and voluntary initiatives may play an important role, they can achieve a target such as that of Kyoto only if firms and households are willing to incur substantial short-run costs in order to further abate their GHG emissions. Third, if the costs are significant, then a macroeconomic analysis is also required to assess their potential national, sectoral, and regional impacts. Our policies below especially emphasize sectors shown by our study to have the most cost-effective options for GHG emission abatement: electricity, transportation, infrastructure, fossil fuel exploitation, production, and transport.

GHG policy should be realistic about consumer preferences if it is to succeed in terms of public acceptability. All too often cost estimates ignore consumer preferences, the assumption being that a limited, financial cost analysis reveals the relative merits to consumers of different technology options. This assumption can lead to a situation, for example, in which the financial cost analysis shows that many people will switch to public transit in the face of higher charges for GHG emissions when the more likely outcome is a dramatic switch to GHG-free private vehicles.

GHG policy should also be realistic about the relative long-term importance of value and preference changes versus financial incentives and technology changes. While value and preference changes are essential, even to provide the support for more effective policies, they are unlikely to be sustainable without key changes in the financial incentives for firms and households. The gains from voluntary initiatives need to be consolidated with policies that financially reward this preference change and with technological evolution that minimizes the dependence of successful policy on continuous vigilance.

Several of the above criteria are consistent with the argument that GHG policy should, wherever possible, seek synergies with other non-GHG values and objectives. An important case is urban livability, where urban land use and infrastructure can be shaped in ways that are consistent with the public's urban lifestyle preferences while making substantial low-cost environmental gains. Thus, the objective is not to push urban residents into densities and locations that they do not desire but to support their goals for safety, reduced congestion, cleaner air, and affordable living in ways that also reduce GHG emissions. The relative importance of these other objectives will play a key role in determining the success of different GHG abatement actions.

Setting GHG policy is a classic case of decision making under uncertainty, and this reality should be embraced instead of ignored or used as an excuse for inaction. This means that policies should be selected based on how well they perform (their robustness) under highly variable outcomes and even highly variable reference cases. How might the economy evolve? What kind of international agreement might eventually follow or replace the one reached at Kyoto? What will the United States do, and how will its economy be affected? How will the pace and character of technological evolution change? Policies must be well positioned to incorporate unexpected technologies, to adapt to shifting targets, and to anticipate and mesh with international policy instruments. Because of this latter consideration, our policies below emphasize market-oriented trading mechanisms over taxes and command-and-control regulations. For example, a GHG permit-trading policy offers the best prospects for links to international permit-trading mechanisms that may be established in the future.

Finally, in the Canadian federal system, policy design must be especially sensitive to regional cost incidence. Federal-provincial cooperation is essential to the development of a coherent national GHG abatement effort, and this means that no region should be asked to bear a disproportionately large share of the burden. Where concentrated impacts could arise, policy design can play an important mitigative role. For example, tradable permits can be mostly grandfathered, meaning that those with high initial emissions are accorded a higher right to emit. GHG tax revenue can also be recycled back to those who initially paid the most while retaining a financial signal that ensures continued efforts by all to reduce GHG emissions. Sector- and technology-specific regulations and subsidies can also be targeted in ways that offset what would otherwise be larger burdens on particular sectors and regions, thus reducing the macroeconomic impacts.

Preferred Policies

Our proposed strategy for Canada is to apply a package of policy instruments that best meets the multiple criteria outlined above. While one often hears of the great potential of voluntarism, on the one hand, and the simplicity and effectiveness of a single, economy-wide policy instrument (e.g., an economy-wide GHG emission tax or a single GHG emission cap with tradable permits), on the other, we believe that the first approach overlooks the importance of incentives in achieving sustained changes in preferences and technologies, while the second overlooks significant implementation barriers.

We have constructed a detailed policy package that would approximate the reductions and cost estimates of our national target simulation. Costs could be slightly higher to reflect the real-world administrative imperfections of trying to coordinate an array of sector-specific policies while pursuing the general goal of the equi-marginal principle. However, the policy package could reduce our estimated cost if it succeeds in accelerating the development of new technologies and the shifts in consumer preferences that would lead to a more rapid commercialization and consequent cost decrease.

While our policy package is somewhat focused on policies that impact a few key sectors, such as electricity, transportation, and municipal transit infrastructure, the complete package must be constructed to ensure that low-cost opportunities in every sector are not missed, especially ones that will be durable over the long term. Therefore, voluntary initiatives, tax credits, and some command-and-control technology regulations should be targeted toward specific low-cost opportunities in industrial equipment and processes, appliances, buildings, landfill gas recovery, transportation, and the other actions identified in our study.

An Enthusiastic yet Sober Approach to Voluntarism

We do not want to underestimate the potential for voluntarism, but we are skeptical that firms and households will voluntarily take on the magnitude of costs and preference changes shown by our study to be necessary. Supporting policies will be required. Governments should continue with their information campaigns and voluntary programs but should clearly and consistently signal the need to go further with careful tax credits, modest financial disincentives, and market-oriented regulations. By being modest initially, such policies allow for adjustment as new information arises about the science of climate change, the pace and character of technological evolution, the actions of other countries, and the state of international negotiations. In this sense, this approach lays the groundwork for future policies that consolidate some of the innovations and preference changes that begin as voluntary initiatives.

Selected Command-and-Control Regulations

Conventional command-and-control regulations can serve as follow-up policies, but they will fail politically if they are applied in an attempt to drive cutting-edge technological or preference change. As in the past, stakeholder consultations can be developed to tighten gradually the minimum energy efficiency standards that already apply to building shells, heating and ventilation systems, lighting, appliances, electronic equipment, vehicles, motors, and industrial equipment. The intent of such regulations should be to eliminate the least energy-efficient, most polluting types of equipment (affecting, say, 25% of sales), not to force on consumers and businesses a single, GHG-optimal type of equipment. Because more stringent restrictions on choice will generate a backlash, we see a fairly modest role for command-and-control regulations in the shift to a less GHG-intensive trajectory.

Subsidies to Support Technologies, Buildings, and Infrastructure, Especially via Tax Credits

Tax credit policies are often criticized by economists because they may involve governments in selecting the winning and losing technologies; do not result in prices that reflect pollution costs, thereby missing the incentive benefits of pollution taxes; and require that undesirable taxes be higher than they otherwise need to be in order to offset the resulting lost government tax revenue. We agree with these concerns but note that tax credits score well on public acceptability and can be effective if designed carefully and with an understanding of relative costs in different sectors and activities in the economy. However, this approach is limited to periods of budgetary surplus, when governments need not increase other taxes in order to provide tax credits, infrastructure grants, and revolving funds for low-interest loans. Tax credits and grants should be designed to minimize government's role in picking technologies by being more performance based. An example would be a tax credit for any electric motor that met a specific set of energy efficiency performance criteria or a tax credit for all equipment that used solar energy directly for water heating or electricity generation. Another example would be infrastructure grants that were not overly specific on how infrastructure might reduce GHG emissions, as long as such reductions could be verified. A third example is to support various types of research and development associated with a CO_2 emission-free energy system, even one that retains fossil fuels as its primary energy source as long as GHG emissions are reduced. Tax credits can be a valuable policy if they are designed to work effectively in concert with the other policies proposed here.

Sector-Specific, Market-Oriented Regulations to Drive Fundamental Change

Sector- and technology-specific policies risk uneconomic outcomes to the extent that they may cause some unnecessarily high-cost actions to be taken. However, a target in the range of the Kyoto Protocol is so far below the level needed just to stabilize atmospheric GHG concentrations that small movements up the marginal cost curves in almost any sector of the economy are likely to be economic in terms of the long-term environmental objective. Therefore, sector-specific policies can be applied in a cautious and flexible manner that minimizes the risk of uneconomic outcomes. Given their relative attractiveness in terms of public and industry acceptability, sector-specific, market-oriented regulations – such as the cap and tradable permit, the RPS, and the VES – can play an important early role in driving fundamental change in technologies and preferences. This kind of policy instrument can alert manufacturers to long-term policy objectives yet give them enough time to design both the technologies and effective marketing strategies. While the short-term impact on emissions may be negligible, the long-term effect can be profound. Some analysts argue that the California VES has played a significant role in motivating a technological revolution in automobile design that will soon cause just as profound a revolution in consumer preferences for vehicles. While Canada may not have the same global influence as California, its internal market is about the same size. In any case, the cost to Canada of simply matching California's standards, as some states have done, is small. Canada could also implement a GHG cap and tradable permit policy for its electricity industry. If this policy is pursued, then it can be employed simultaneously with RPS or as a substitute for RPS.

A Modest Economy-Wide Cap and Tradable Permit System that Operates Initially like a Tax

The Canadian public is not yet ready for environmental tax shifting to play the leading role in GHG emission abatement. Likewise, the singular use of an aggressive cap and tradable permit system, as assumed in our cost analysis, will fail the political acceptability test. However, initiatives by industry would be encouraged by at least some indication from government that GHG emissions will one day have a cost. A modest, economy-wide cap and tradable permit system could be established fairly quickly if it were structured with an upper cost limit in its early years.[35] There would be a price cap on emissions permits, thereby assuring everyone that Canada was not about to embark on a potentially costly exercise. A specific policy design might look like this. The federal government would allocate a percentage (say 70% to 90%) of emission permits based on emissions in a base year (grandfathering) and then provide an unlimited number of additional permits at a low fixed price of $10/t CO_2e. With this mechanism, the average cost of GHG emissions would be lower than $10, a function of the percentage of emissions covered by free, grandfathered permits. However, the marginal cost of GHG emissions would be $10/t CO_2e, a useful financial signal for embryonic industry permit-trading programs. The fee for additional permits would be like a tax in that additional payments accrue to government. To avoid the accusation of raising taxes, government would need to develop a mechanism to return the modest GHG permit revenues to the regions and sectors most affected or perhaps into GHG-related research and development programs.[36] While this type of policy on its own would not enable the country to achieve a target such as the Kyoto Protocol, it would complement other policies and possibly contribute to changing awareness by firms and households, a change that would lay the groundwork for future, stronger policies as public attitudes evolve and as more is learned about abatement options, climate change risks, and international agreements. Laying the foundation of a cap and tradable permit system will also position Canada to quickly take advantage of future international permit-trading mechanisms.

Conclusion

Of the range of policy options for GHG emission abatement, governments today are especially enamoured with noncompulsory approaches such as information programs, voluntary initiatives, public-private partnerships, and modest public subsidies in the form of tax credits. However, our cost estimates suggest that these approaches alone will not be enough if Canada is to achieve a GHG abatement target such as that of Kyoto. Some integration of more compulsory policies will be required. The challenge is that such policies generally involve a significant increase in energy prices, and this increase will quickly provoke a reaction in consumers and the media that politicians will be forced to respond to.

Fortunately, there may be another way of providing long-term signals to technological innovation and product marketing. A new generation of market-oriented regulatory policies can provide a significant, long run financial signal without substantially increasing energy costs and budgets in the short term. Early implementation of such policies can help to provide information about technological opportunities that in turn can reduce major cost uncertainties.

This approach to policy making is based on our conclusion to Chapter 4, where we stated that the costs of GHG emission abatement are both high and low depending on the perspective. While the total long-run costs to society may not be that high, they can appear high when the best available means of motivating actions is via dramatic increases in prices, with significant short-run budgetary and political impacts. The challenge for policy making is to understand and work with this duality.

8
The Next Steps: Addressing the Uncertainties of GHG Abatement Costs

Summary of Method and Findings

Making a decision on mitigating GHG emissions requires several cost considerations. What will different amounts of GHG emission reduction cost? How will these costs be distributed among sectors of the economy and regions of the country? How does the choice of policy instrument affect the magnitude of costs and their distribution among these sectors and regions? How will action or inaction by our key trading partners and competitors affect our costs? What are the nonclimate benefits from actions to reduce GHG emissions, and how should they be included in a cost analysis? How significant are the uncertainties associated with cost estimates? How can these costs be compared with the risks of climate change disruption in choosing a target and acting on it?

For well over a decade, the cost of reducing GHG emissions has been the subject of heated debate, with cost estimates scattered over a wide range. Some researchers argue that reduction of GHG emissions involves capitalizing on many profitable opportunities that in aggregate can stimulate economic growth. Others argue that the costs will be significant, with a stagnating effect on economic growth and a potentially devastating effect on some sectors and regions.

Divergent views on the costs of GHG emission reduction are explained in part by differing views about the significance of and threat from human impacts on the environment. Those who worry about pervasive environmental deterioration have a tendency to see the costs of emission reduction as small, while those who worry about overzealous environmentalism obstructing economic growth have a tendency to see the costs as large. But among climate policy researchers, the divergent cost estimates are in large part the result of two different methodological approaches to modelling technological change.

Our approach argues that the preferences of consumers, as expressed by their willingness to pay more for a particular commodity in spite of the availability of apparent substitutes, must be reflected in cost estimates. Thus, the cost of GHG emission abatement is not just the relative differences in financial costs (capital and operating) of competing equipment but also the cost of overcoming consumer preferences for certain technologies over others. To most economists this is the proper way to estimate the cost of reducing GHG emissions. With preferences included, their models have tended to show high costs of GHG emission abatement.

However, what this approach can miss is the potential for dramatic technological innovations to change the technical potential and relative financial costs of lower emission alternatives as well as to influence the preferences of consumers and businesses. For example, a vigorous drive for technological innovation in vehicle design could significantly reduce the relative financial costs of less polluting vehicles. This technological innovation could also change consumer preferences for reasons other than just financial costs, perhaps because the cleaner vehicle can be successfully marketed in terms of environmental friendliness. The resulting shift in preferences would further reduce the cost estimate of GHG abatement actions.

Also, when it comes to GHG mitigation policy, many factors hinder policy makers from implementing a simple, economy-wide instrument such as a GHG tax or a cap and tradable permit system. Likely policies, some of which are already in place, tend to be technology- and sector-specific tax credits, direct subsidies, command-and-control regulations, and market-oriented regulatory instruments. An aggregate representation of the energy system and the economy – the norm for this approach – is unable to help the policy maker in estimating the likely cost effectiveness and GHG emission reduction contribution of such packages of technology-specific policies.

An alternative approach, applied more frequently by engineers, physicists, and environmental advocates, focuses on technologies that produce or consume energy in order to estimate how technological innovation and turnover of the existing equipment stock can lead to different quantities of total energy use, relative energy shares, and emission levels. This approach examines the individual production and consumption services provided by technologies (process heat, space heat, lighting, mobility, etc.) and compares the relative financial costs of alternative technologies in satisfying a service. Technologies that are eligible to provide the same service are often assumed to be perfect substitutes in all respects other than financial costs. The results of these models suggest that, if low-emission technologies were widely adopted, GHG emission reduction would be achieved at low cost.

But some technologies, in spite of apparently providing the same service, will differ in the view of businesses or consumers in terms of important attributes such as perceived risk and quality of service. If we used a GHG tax – or the permit price resulting from a GHG permit-trading system – to push consumers and businesses toward the less-desired technology, we would then have an estimate of what permit price or tax would be required to propel the entire economy toward a particular GHG emission reduction target. If this tax or permit price is used for estimating the cost of GHG emission reduction, then one can see that the cost estimate will be much higher than the pure financial cost issuing from the technology-focused approach.

The method that we apply to cost estimation acknowledges the legitimacy of the criticisms of both approaches, and thus technologies and preferences are both integral to our cost estimation methodology. Because it is technology explicit, our costing model, CIMS, uses the same technology-specific financial cost information as technology-focused models. Because it is also preference sensitive, CIMS requires estimates of the intangible values that consumers and businesses associate with specific technologies as well as the risk aversion that both may exhibit in certain types of decisions, such as energy efficiency investments that have lengthy payback periods. In this sense CIMS is a predictive model. It attempts to indicate the market prices, technology-specific subsidies, or regulations (technology or sector specific) that would be required to achieve a particular target of GHG emission reduction. In the case of government implementation of an economy-wide cap with a tradable permit system, CIMS estimates what the market price of the tradable permits would be in order for Canada to satisfy the emission cap. Over a two-year period, as consultants to the National Climate Change Process and in separate work, we have applied CIMS to estimate, at a detailed level, the national, sectoral, and regional costs of Canada initiating a substantial policy thrust to achieve significant GHG emission reductions. According to our simulations, if Canada had applied an economy-wide cap and tradable permit system in 2000 as its solo policy to reach its Kyoto target by 2010, then the resulting tradable permit price would be in the range of \$120/t CO_2e. This amount translates into significant increases in final energy prices. When the prices and investments are fed into a macroeconomic model, we estimate a 3% reduction in Canada's cumulative GDP growth relative to the reference case. This means that the economy might grow by 27% over the 10 years from 2000 to 2010 instead of by 30%.

We are aware of only a few models similar to CIMS, although international interest in this type of modelling has risen dramatically in recent years. The principal model of the US government, NEMS, is very similar. Although the model has not been applied to Canada, its application to the United States provides a contrast with our cost estimates for Canada. The US emission target is comparable to Canada's (7% below 1990 levels as opposed to 6% below), and both countries have somewhat similar economies, urban form, and infrastructure. In a recent exercise, NEMS was used to estimate the cost to the United States of achieving its Kyoto target entirely by domestic actions. As in our Canadian application, a percentage of the target was achieved by forestry and agricultural actions and was not directly modelled in NEMS, while the rest was achieved by energy-related actions. From the NEMS simulation, the cost of a carbon tax in the United States would be $294 (US)/tonne of carbon. When this amount is converted to Canadian currency, and from carbon to CO_2e, it is close to our estimated value of $120/t CO_2e.

In general our cost estimates suggest that Canada needs more than a voluntaristic approach to make the fundamental reductions in GHG emissions necessary to reach a target such as that of Kyoto. The costs of significant reductions – and the energy price increases needed to motivate action – are substantial. Past evidence provides no support for the suggestion that businesses or consumers are willing to incur significant costs on a voluntary basis. This is not to suggest that voluntarism should be abandoned as a policy thrust but simply that stronger policies will also be required.

We have produced a cost curve that shows how the GHG emission permit-trading price must change in order to achieve different national target levels. Such a curve is useful in several respects. First, it provides information to the government on low-cost options and therefore on what it might cost for Canada to take additional unilateral steps to reduce GHG emissions. Second, Canada can now participate in international negotiations with a much better sense of what different target levels might mean to the country's economy. The detailed sectoral and regional estimates from the simulation of the Kyoto target can help in designing policies that minimize the cost impacts for different sectors and regions. These estimates also provide information that is needed for negotiations between the federal and provincial governments on how to allocate domestically the burden of Canada's target for GHG emission reduction.

However, the research conducted thus far is preliminary. Follow-up research should test different allocation mechanisms for their regional and sectoral impacts. This research is now under way both within the National Climate Change Process and in our independent research and that of others.

Technology and Preference Uncertainties and Their Impacts on Cost Estimates

Whether we are talking about the Kyoto target or the broader range of GHG emission reductions implied by our cost curve, how confident can decision makers and the public be about these cost estimates? While we cannot have certainty about the future, there are some developments that we can be more confident about than others. For example, we know that urban form will not change profoundly over the next 10 years, and even over the next 20 years there will be considerable inertia. We can use the structure provided by a model such as CIMS to identify and probe this and other rates of change and thus identify major sources of uncertainty. This application of a model is referred to as conditional analysis: an analysis that explores the effects on our cost estimates of different conditions for key parameters or input assumptions.

Conditional analysis in this application can be broken into two dimensions. One dimension is to conduct sensitivity analysis in order to determine which parameters and input assumptions have significant impacts on the results. Alternative values for parameters and inputs are simulated with the model. Where the significance of these changes for the model outcome is small, uncertainty about the future value is assumed to be of less consequence. The other dimension is to try to determine in advance (ex ante) those parameters and input assumptions for which there is currently a high degree of uncertainty for the time period in question. This uncertainty is ascertained by an analysis of technologies and preferences. For example, the future costs and technical characteristics of the technologies and geological formations that might capture and sequester GHG emissions from fossil fuels are today highly uncertain.

The combination of these two dimensions leads to four possible conditions, as shown in the simple matrix of Figure 8.1. Of great concern are parameter and input assumptions that are both highly uncertain and of high significance to the simulation outcomes, as depicted by the bottom right cell in the table. These assumptions should have the highest priority for further analysis and, as we explain below, for policies that provide an opportunity to learn more.

In the results chapters, we discussed some of the sensitivity and uncertainty analysis that has been conducted thus far. This analysis includes sensitivity runs as well as a comparison of our results with those of other key models. Much remains to be explored. In general we can characterize the uncertainty from this research into two broad categories: (1) the direction

Figure 8.1

Uncertainty and significance of parameters and inputs

	Low significance to outcome (from sensitivity analysis)	**High significance to outcome** (from sensitivity analysis)
Low degree of uncertainty (ex ante evaluation)	Lower research concern	Lower research concern
High degree of uncertainty (ex ante evaluation)	Lower research concern	Highest priority for further analysis and learning policies

and pace of key technological changes, and (2) the potential for change in consumer and producer preferences. We discuss each in turn.

An examination of actions to reduce GHG emissions suggests that there are a few key technological potentials that can have a dramatic effect on costs. One that stands out in our view is the emerging prospect for transforming fossil fuels into GHG-clean energy. Will it be technically possible and economically attractive to transform carbonaceous fuels in large quantities into forms of energy that, in concert with key technologies, will have almost zero emissions throughout their life cycles? In the horizon to 2030, we have modelled only one technology that separates CO_2 from post-combustion flue gases of coal-based electricity generation. However, researchers in this area are currently exploring a more dramatic transformation of the energy system that could have a significant impact by 2030 and possibly even by 2020. In this scenario existing and emerging technologies would enable the conversion of all forms of fossil fuels to synthetic gas and eventually hydrogen and electricity with the potentially harmful emissions separated precombustion from the source fuel and sequestered underground in old oil and gas reservoirs, coal beds, and deep saline aquifers. If this option progresses as some analysts are predicting, then the cost curve for reducing GHG emissions would become flatter than we have estimated. In the future we will use our model to test technological evolution such as this for its effects on the aggregate, regional, and sectoral costs of GHG emission reduction in Canada.

The second major area of uncertainty is consumer preferences. In this study we have shown that our explicit accounting for consumers' value losses explains much of the difference in cost estimates between our model and those generated by a bottom-up, financial optimization model. Economists generally treat consumer preferences as stable because, first, economists tend to focus on time periods during which a dramatic shift in preferences is unlikely and, second, the direction of preference shift is highly uncertain. In effect, one can postulate any change in preference to justify any cost estimate. Thus, economists are skeptical about cost estimates in which the analyst conveniently postulates that consumers will suddenly find certain products or technologies to be perfect substitutes in spite of historical evidence to the contrary.

However, with a long-term, fundamental challenge such as climate change, it is equally problematic to treat preferences as static. While the Kyoto agreement provides a convenient near-term target to focus analysis and policy, it is but a small step along the path to stabilization of atmospheric concentrations of GHGs. Much more dramatic transformation is required if society is to reach such a long-term goal. This transformation will undoubtedly require profound technological changes. But it will also require shifts in preferences toward technologies and lifestyles that reduce emissions. Indeed, this is currently the presumption behind the government's policy thrust in favour of voluntary initiatives. In Chapter 7 we discussed the types of policies that might contribute to shifting preferences. We especially noted policies that mobilize manufacturers to market cleaner technologies by concentrating on consumer and business preferences. Some of the initiatives of the electric utility industry in the 1980s and 1990s provide a useful example; they attempted to generate a market transformation by achieving critical penetration thresholds for high-efficiency appliances in the household sector and high-efficiency motor systems in the industrial sector. A more recent example is the vehicle emission standard initiated in California that motivates vehicle manufacturers to market cleaner vehicles: the better manufacturers become at inducing consumers to pay extra for such vehicles, the lower industry's net cost of compliance with the standard. This approach implies that consumers, once their preferences have shifted, suffer less loss of value than originally calculated, lowering the estimated total social cost of reducing GHG emissions. In other words, the market price of tradable permits would be lower for the same target simply because consumers and businesses would not need such a large compensation or penalty to choose lower-emission technologies and lifestyles.

In predicting shifts in consumer preferences, there are promising research avenues to pursue. For example, we and other researchers are currently assessing the extent to which consumers perceive private vehicles and public transit as substitutes. If they can be seen as close substitutes for a service such as commuting, then it may be relatively easy to motivate a shift from private vehicles toward public transit as an emission reduction strategy. If they are not seen as close substitutes, then consumers may be willing to pay a great deal more for cleaner cars (electric, hydrogen, ethanol) before they are willing to shift in large numbers toward public transit. Our sources for estimating the tendency and future potential for change of consumer preferences include both revealed preference studies – historical data on vehicle use and public transit taken from a cross-section of locations – and stated preference studies – surveys that ask people what they would do when faced with different choices of more expensive clean cars compared to public transit.

While this research will help to improve our understanding of preferences and our predictive abilities, the long-term evolution of preferences will remain highly uncertain. Therefore, we also intend to intensify our use of CIMS for conditional analysis of these critical uncertainties. Some of the research reported here provides an indication of this application of the model. In particular we showed in Chapter 4 how a policy such as the vehicle emissions standard might dramatically change vehicle marketing strategies, with consequences for consumer preferences and the costs of GHG emission reduction. This highly speculative exercise gives the model user an understanding of the importance of this particular technology choice, and this particular sector of the economy, for the aggregate cost estimates. We intend to conduct similar studies for buildings, appliances, various types of industrial and commercial equipment, and even specific urban form decisions related to the integration of commercial and residential activity with expansion of public transit.

In summary, our initial modelling of GHG emission reduction options has helped to identify factors that are highly uncertain and to which the cost outcome is highly sensitive. Conditional analysis can help to probe these sources of uncertainty in order to better inform policy makers of the range of possible outcomes in terms of what is achieved and at what cost.

However, a long-term problem such as GHG emission reduction also provides an opportunity over time to learn more about these areas of great uncertainty. Surveys can provide some indication. But the outcome of such research will remain highly conjectural as long as the situation being considered is hypothetical. An important way for learning about any dynamic system is to initiate early changes to the signals that drive the system, even if only on a tentative and experimental basis. There may be a substantial knowledge-acquiring value from early, modest policies to reduce GHG emissions.

Policies to Reduce Technology and Preference Uncertainties

Our goal in this study is to develop and apply the most appropriate method for providing decision makers with cost information so that they can make decisions about how and by how much to reduce GHG emissions. This test application is to Canada, but the methodological approach has universal applicability. We have estimated costs for the Kyoto target. We have differentiated costs in terms of a policy target (sectoral targets versus a single national target) and in terms of regional and sectoral impacts. We have also shown how these costs can differ if the target changes.

Our study does not enable us to comment on the correct target for Canada alone because we have ignored the benefits side of the issue; this would be a very different exercise involving the estimation of global benefits and global costs within a broad decision analysis framework. However, our analysis of uncertainty suggests that Canada could take actions today that may provide substantial benefits in terms of improving our information about the costs of key technology prospects and the potential for preference shifting by consumers and businesses. Moreover, if one accepts the argument that international agreements to reduce anthropogenic GHG emissions, in some guise, will eventually be ratified and implemented, then these immediate actions are justified even without a current commitment to action by other important countries such as the United States.[1]

There is a value to information about the eventual cost of key technology- and preference-related options for GHG emission reduction, and policies enacted today may be able to provide more of this valuable information in the near future at low cost. This information may in turn improve the policy choices made in future years and thereby reduce the total cost to society of reducing GHG emissions.[2]

Several of the policies in our policy package in Chapter 7 fit this characterization. For example, the market response to modest levels of technology-specific tax credits will inform policy makers of the relative preference for some of the technologies that reduce GHG emissions. This response can feed back to cost estimates and policy choices. Likewise, market-oriented regulations such as the RPS and the VES provide long-run cost signals to producers and consumers without having dramatic effects on the current average prices of electricity and cars. Again, at low current costs, these policies can provide improved technology cost and consumer preference information, which in turn reduce costing uncertainties and policy design problems.

Are the Costs High or Low?

Our method and analysis suggest that the costs of Canada meeting a Kyoto target can be seen as large or small depending on one's perspective. On the one hand, attaining a dramatic reduction in emissions between 2000 and 2010 may require substantial increases in energy prices. Given the lack of tolerance for price increases, the costs will seem to be large both economically and politically. Also, because preferences and technological options are more static and limited in the short run, the potential for adjustment is more limited and the costs, including lost consumers' value, that much higher. On the other hand, energy costs are less than 5% of the budget of a typical household and the cost of production of all but the most energy-intensive industries. Even a substantial energy price increase does not represent a large drop in real income, especially in terms of strict financial flows. Consumers and businesses will buy equipment and make other adjustments that, after the initial investment costs, will reduce overall energy costs. As a percentage of GDP, the overall costs may not be that great, as our results show. Indeed, our cost estimates may be too high, if long-term technological innovation and preference shifting reduce financial costs and lost consumers' value.

Our analysis therefore drives us to the following conclusions and suggestions. Because some GHG emission reduction actions can start with small initial costs, while providing critical long-term signals that should provoke changes with larger long-term effects, Canada can start now with modest unilateral policies without risking substantial economic repercussions. The focus should be on key technology- and preference-shifting strategies that are low cost and perhaps of small effect initially but can lead to technology and preference transformation over the long term. Also, these modest steps will reduce uncertainty about the nature of consumer preferences and technologies by providing critical real-world evidence of the cost implications of technology innovations and efforts to transform market preferences. These actions, taken today, will better position Canada to make more substantial, better-informed, and less uncertain policy initiatives in the future, presenting the prospect of more confident policy making and more substantial cost savings over the long term.

Notes

Introduction

1 White House press secretary Ari Fleischer, 8 May 2001, responding specifically to whether President Bush would call on drivers to reduce their fuel consumption, <http://www.eenews.net>.

2 A poll conducted by Pollara shows that Canadians support significant action to combat global warming, although they make such statements without a clear idea of what this might mean in terms of personal costs. The poll showed that 83% of respondents believe that severe weather events are related to climate change. "Canadians Fear Effects of Climate Change: Poll," *Ottawa Citizen*, 7 August 2000.

3 "E.U. Mulls Plan after U.S. Kyoto Snub," *New York Times,* 5 April 2001.

4 At the time the government could not have been aware of the extent to which sustained economic growth in the 1990s would result in such a large reduction requirement.

5 Address to the National Forum on Climate Change, National Roundtable on Environment and Economy, Ottawa, 17 February 1998.

6 Joshua Avram, "Will Canada Choke on Kyoto Fumes? Westerners Fearfully Assess the Likely Costs of the Latest Eco-Conference," *Alberta Report* 24, 48 (1997): 18. Klein's environment minister Ty Lund went so far as to raise the possibility of western separation when he warned that his province would not submit to the national target without a fight. "Premier Klein Overstates Kyoto Climate Deal's Impact on Alta Economy: Experts," Canadian Press Newswire, 23 July 2001, on-line database available from Canadian Business and Current Affairs.

7 The national process and our role in it are explained in Analysis and Modelling Group, National Climate Change Process, *An Assessment of the Economic and Environmental Implications for Canada of the Kyoto Protocol* (Ottawa: National Climate Change Process, 2000) and in our report *Integration of GHG Emission Reduction Options Using CIMS* (Vancouver: Energy Research Group/M.K. Jaccard and Associates, 2000).

Chapter 1: The Climate Change Threat

1 Venus's mean temperature is 452°C, while Mars shows average temperatures of -63°C. M.H. Hart, "Habitable Zones about Main Sequence Stars," *Icarus* 37 (1979): 351-57.

2 There is no easily defined "end" to the atmosphere. Atmospheric scientists say that it extends 1,600 km from the Earth's surface, but the first 160 km contain 99% of the atmosphere's mass. The atmosphere is less transparent to some forms of electromagnetic radiation than others. Electromagnetic radiation refers to a broad spectrum of electromagnetic waves of varying wavelength and frequency, each with its own characteristics and energy content.

3 By slowing the rate of heat loss, this retention raises the Earth's surface temperature. But the eventual heat loss from the Earth equals the heat gain from the Sun; otherwise, the Earth would not remain at a stable temperature but continue to heat up.

4 Ozone, for example, absorbs ultraviolet light, a form of electromagnetic radiation of higher frequency and shorter wavelength than visible light, which initiates chemical reactions in organic material.
5 Aerosols – suspended solid or liquid particles in the atmosphere (dust, smoke, sea salt, or organic materials) – retain infrared radiation. They tend to be short-lived in the atmosphere but are constantly replenished by both natural biologic activity and anthropogenic activity, including the combustion of fossil fuels and even driving vehicles on dusty roads. Extended use of fossil fuels, especially those that generate soot or smoke, has increased the atmospheric levels of aerosols. However, little information is available concerning estimates of future concentrations in the atmosphere. It depends on how societies utilize carbon fuels in the future. IPCC scenarios suggest that natural aerosols will counteract to some extent the global warming effect.
6 Intergovernmental Panel on Climate Change, *IPCC Second Assessment Report: Climate Change 1995* (Cambridge: Cambridge University Press, 1996); Intergovernmental Panel on Climate Change, *Climate Change 2001: Third Assessment Report* (Cambridge: Cambridge University Press, 2001).
7 A chapter is devoted solely to the issue of human health impacts in the IPCC's third assessment report. See *Climate Change: Impacts, Adaptation, and Vulnerability,* vol. 2 of IPCC, *Climate Change 2001.*
8 IPCC, *IPCC Second Assessment Report,* 33.
9 Ibid.
10 For example, tonnes of paper produced may remain the same between two years while the market price increases by 6% and the inflation adjuster applied to this sector of the economy shows only a 4% increase. This gap between the inflation adjuster and the actual price change translates into an apparent 2% increase in output, which would lower the energy intensity of output. In reality, neither the technical energy efficiency of production nor the physical level of output has changed.
11 Some analysts argue that societies go through an inevitable evolution toward a service- and information-oriented economy, with consequent reduced use of materials, and that this development alone can dramatically reduce GHG emissions. To explore this possibility, the equation can be further disaggregated by reporting the E/Q for each sector of the economy and then summing them up. M. Bernstam reports a strong correlation between increasing economic development and decreasing levels of pollution. See M. Bernstam, *The Wealth of Nations and the Environment* (London: Institute of Economic Affairs, 1991). Some researchers refer to this as the *Environmental Kuznets Curve*. However, there are many analysts who argue that, while economic development is associated with a more efficient use of resources, more stringent pollution control standards, and better pollution control equipment, flows of material and energy continue to increase, ultimately with negative environmental consequences. They may surface in forms such as accumulations of persistent toxins in air, water, and the biosphere, loss of biodiversity, and climate disruption. For diverse views, see D.S. Rothman and S.M. de Bruyn, eds., *Probing into the Environmental Kuznets Curve Hypothesis,* special issue of *Ecological Economics* 25 (1998).
12 El Niño, La Niña, and changes in the oceans' thermohaline circulation (i.e., transportation of heat along with water in large ocean currents) provide examples of what increases in oceanic temperature may lead to.
13 With 1.7 trillion barrels in the Athabasca region in Alberta, the oil sands contain more oil than all the known reserves of Saudi Arabia.
14 Coal is by far the most plentiful of fossil fuels, with known reserves large enough to supply present rates of consumption for over 450 years. If one considers the total resource, including coal not presently economical, supply could last an estimated 1,000 to 1,500 years. United Nations Development Programme, *World Energy Assessment: Energy and the Challenge of Sustainability* (New York: United Nations, 2000), 6.
15 Annex 1 refers to signatory countries that have made GHG reduction commitments.
16 In recognition of countries' different economic circumstances, energy requirements, and differing capacities to make emission reductions, each country has a specific, differentiated target. Some countries are entitled to increase their emissions from the 1990 baseline,

including Iceland, Norway, and Australia. Australia's target reflects its strong production focus in GHG-intensive industries and limited opportunities to switch to less GHG-intensive forms of energy.

17 Leakage refers to the possible shift of activity, primarily industrial, from countries under some GHG constraint to countries where no such constraint exists. Not only is this a competitiveness issue, but also the developing country is not as likely to possess the infrastructure or technology to produce goods as efficiently, generating the various "leaked" products at higher levels of CO_2 than had production remained where it was.

18 D.G. Victor, *The Collapse of the Kyoto Protocol and the Struggle to Slow Global Warming* (Princeton: Princeton University Press, 2001).

19 While there is uncertainty about the appropriate concentration limits to mitigate climate change impacts, the European Union has suggested that a doubling of concentrations (550 parts per million, by volume) should constitute the upper limit of CO_2 concentrations. To achieve this objective, global GHG emissions would have to return to their 1990 levels by the year 2100. Beyond 2100, emissions would need to decrease further and stabilize at less than 2 billion tonnes of carbon, a 75% decrease from 1990 emission levels. The Kyoto Protocol represents a 5% decrease in emissions in CO_2 from 1990 levels for industrialized countries only. See R. Priddle, "The Energy Dimension of Climate Change," Parliamentary Group for Energy Studies, US Energy Information Administration, 4 June 1997.

20 Heather Smith, *Canadian Federalism and International Environmental Policy Making: The Case of Climate Change* (Kingston: Institute of Intergovernmental Relations, Queen's University, 1998).

21 The NCCP includes participation from federal and provincial energy and environment ministers, the prime minister, provincial premiers, and territorial leaders.

22 National Climate Change Process, *Canada's First National Climate Change Business Plan* (Ottawa: NCCP, 2000).

23 See the Greenhouse Gas Emission Reduction Trading Pilot's Web site at <http://www.gert.org>.

24 These reports are available on their respective Web sites: Canadian Chemical Producers' Association, <http://www.ccpa.ca>; Canadian Petroleum Products Institute, <http://www.cppi.ca>.

25 M. Bramley, *Greenhouse Gas Emissions from Industrial Companies in Canada: 1998* (Drayton Valley, AB: Pembina Institute, 2000).

26 This group is a subcommittee of the National Air Issues Steering Committee, an intergovernmental committee that responds to broader air issues.

27 These are available from NCCP's Web site, <http://www.nccp.ca>.

Chapter 2: The Challenges of Estimating Emission Reduction Costs

1 See discussion in, for example, R. Boardman, ed., *Canadian Environmental Policy: Ecosystems, Politics, and Process* (Toronto: Oxford University Press, 1992); and M. Hessing and M. Howlett, *Canadian Natural Resource and Environmental Policy* (Vancouver: UBC Press, 1997).

2 R. Repetto and D. Austin, *The Cost of Climate Protection: A Guide for the Perplexed* (Washington, DC: World Resources Institute, 1997).

3 J. Robinson, "Of Maps and Territories: The Use and Abuse of Socioeconomic Modelling in Support of Decision-Making," *Technological Forecasting and Social Change* 42, 3 (1992): 147-64.

4 Analysis and Modelling Group, National Climate Change Process, *Canada's Emissions Outlook: An Update* (Ottawa: NCCP, 1999).

5 Although not included in the Kaya Identity here, the reference case also includes forecasts of the evolution of forestry and agricultural practices and related GHG emissions.

6 In spite of this, some argue that, since the world must go far beyond the Kyoto reductions to stabilize GHG concentrations in the atmosphere, government policy should begin now to influence the broader driving forces determining population size, income level, and economic structure.

7 This definition was originally developed by one of the authors and presented to the National Climate Change Process, where it was incorporated into analysis guidelines for the Issue Tables.

8 M. Grubb, J. Edmonds, P. ten Brink, and M. Morrison, "The Cost of Limiting Fossil-Fuel CO_2 Emissions: A Survey and Analysis," *Annual Review of Energy and Environment* 18 (1993): 397-478. M. Jaccard and D. Montgomery, "Costs of Reducing Greenhouse Gas Emissions in the United States and Canada," *Energy Policy* 24, 10-11 (1996): 889-98.

9 Discounting is the reverse of compounding. While there is much debate on the correct value for the social discount rate, it is roughly somewhere between the interest rate on long-term savings and the marginal, pretax rate of return on investments (corrected for different levels of risk and inflation).

10 See M. Brown, M. Levine, J. Romm, A. Rosenfeld, and J. Koomey, "Engineering-Economic Studies of Energy Technologies to Reduce Greenhouse Gas Emissions: Opportunities and Challenges," *Annual Review of Energy and the Environment* 23 (1998): 287-385; A. Lovins and H. Lovins, "Least-Cost Climate Stabilization," *Annual Review of Energy and the Environment* 16 (1991): 433-531; and R. Williams, "Low-Cost Strategies for Coping with CO_2 Limits," *The Energy Journal* 11, 4 (1990): 35-59.

11 For the arguments pro and con, see H. Huntington, L. Schipper, and A. Sanstad, eds., *Markets for Energy Efficiency,* special issue of *Energy Policy* 22, 10 (1994). For the specific arguments by economists, see G. Metcalf, "Economics and Rational Conservation," *Energy Policy* 22, 10 (1994): 819-25; R. Sutherland, "The Economics of Energy Conservation Policy," *Energy Policy* 24, 4 (1996): 361-70; and A. Jaffe and R. Stavins, "Energy-Efficiency Investments and Public Policy," *The Energy Journal* 15, 2 (1994): 43-65. For the arguments in favour of the bottom-up approach, see D. Wilson and J. Swisher, "Exploring the Gap: Top-Down versus Bottom-Up Analysis of the Cost of Mitigating Global Warming," *Energy Policy* 21, 3 (1993): 249-63; A. Sanstad and R. Howarth, "Normal Markets, Market Imperfections, and Energy Efficiency," *Energy Policy* 22, 10 (1994): 811-18.

12 This and following characteristics apply to firms and households, although there is an assumption and evidence that financial considerations more closely represent the key decision factors for firms.

13 Thus, a consumer may still purchase a fridge out of necessity but will not make the incremental capital outlay for a high-efficiency fridge. The benefit of not waiting is sometimes referred to as option value, although there is debate about this and related terms. See R. Pindyck, "Irreversibility, Uncertainty, and Investment," *Journal of Economic Literature* 29, 3 (1991): 1110-52.

14 J. Hausman, "Individual Discount Rates and the Purchase and Utilitization of Energy-Using Durables," *Bell Journal of Economics* 10, 1 (1979): 33-54.

15 See J. Scheraga, "Energy and the Environment: Something New under the Sun?" *Energy Policy* 22, 10 (1994): 798-803. If the compact fluorescent also costs a lot more, then the irreversibility issue of the previous point is also relevant here.

16 See A. Jaffe and R. Stavins, "The Energy-Efficiency Gap: What Does It Mean?" *Energy Policy* 22, 10 (1994): 804-10; and M. Jaccard, J. Nyboer, and A. Fogwill, "How Big Is the Electricity Conservation Potential in Industry?" *The Energy Journal* 14, 2 (1993): 139-56.

17 Economists use the term *elasticity of substitution* (ESUB) for the responsiveness of relative demand to changes in relative prices (e.g., the ESUB between capital and energy reflects energy efficiency investments) and the term *autonomous energy efficiency index* (AEEI) for the rate at which energy efficiency changes even with relative prices held constant.

18 K. Train, *Qualitative Choice Analysis* (Cambridge, MA: MIT Press, 1986); K. Train and T. Atherton, "Rebates, Loans, and Customers' Choice of Appliance Efficiency Level: Combining Stated- and Revealed-Preference Data," *The Energy Journal* 16, 1 (1995): 55-69.

19 Intergovernmental Panel on Climate Change, *IPCC Second Assessment Report: Climate Change 1995* (Cambridge: Cambridge University Press, 1996), Chapters 8 and 9.

20 The well-known joke is that an economist would not pick up a $20 bill from the sidewalk because if it were truly a $20 bill someone would have already done so.

21 See B. Norton, R. Constanza, and R. Bishop, "The Evolution of Preferences: Why 'Sovereign' Preferences May Not Lead to Sustainable Policies and What to Do about It," *Ecological Economics* 24, 2-3 (1998): 193-211; and I. Peters, F. Ackerman, and S. Bernow, "Economic Theory and Climate Change Policy," *Energy Policy* 27, 9 (1999): 501-04.

22 See M. Grubb, "Policy Modelling for Climate Change: The Missing Models," *Energy Policy* 21, 3 (1993): 203-08; M. Jaccard, A. Bailie, and J. Nyboer, "CO_2 Emission Reduction Costs in the Residential Sector: Behavioural Parameters in a Bottom-Up Simulation Model," *The Energy Journal* 17, 4 (1996): 107-34; H. Jacobsen, "Integrating the Bottom-Up and Top-Down Approaches to Energy-Economy Modelling: The Case of Denmark," *Energy Economics* 20, 4 (1998): 443-61; and C. Koopmans and D. te Velde, "Bridging the Energy Efficiency Gap: Using Bottom-Up Information in a Top-Down Energy Demand Model," *Energy Economics* 23, 1 (2001): 57-75.

23 For a technical and economic survey of the available range of energy sources and energy technologies to address climate change and other sustainability concerns, see United Nations Development Programme, *World Energy Assessment: Energy and the Challenge of Sustainability* (New York: UNDP, 2000).

24 M. Jaccard, L. Failing, and T. Berry, "From Equipment to Infrastructure: Community Energy Management and Greenhouse Gas Emission Reduction," *Energy Policy* 25, 11 (1997): 1065-74. For an illustration of a developing-country application, see B. Sadownik and M. Jaccard, "Sustainable Energy and Urban Form in China: The Relevance of Community Energy Management," *Energy Policy* 29, 2 (2001): 55-65.

25 The shapes of such curves are inevitably speculative. See A. Gritsevskyi and N. Nakicenovic, "Modeling Uncertainty of Induced Technological Change," *Energy Policy* 28, 13 (2000): 907-22; and C. Azar and H. Dowlatabadi, "A Review of Technical Change in Assessment of Climate Policy," *Annual Review of Energy and Environment* 24 (1999): 513-44.

26 J. Edmonds, R. Roop, and M. Scott, *Technology and the Economics of Climate Change Policy* (Washington, DC: Pew Center on Global Climate Change, 2000); L. Goulder and S. Schneider, "Induced Technological Change and the Attractiveness of CO_2 Abatement Policies," *Resource and Energy Economics* 21, 3-4 (1999): 211-54; H. Dowlatabadi, "Sensitivity of Climate Change Mitigation Estimates to Assumptions about Technological Change," *Energy Economics* 20, 5-6 (1998): 473-94.

27 Jaccard, Failing, and Berry, "From Equipment to Infrastructure."

28 A significant reason for the lower energy intensities of European cities is that historical and modern urban-planning decisions have led to higher density, greater mix of density uses, and better coordination of urban development with an extensive urban transit system and in many cases district heating systems.

29 There are other dimensions to the hierarchy, but they are not explored here. On a spatial dimension, decisions at the top of the hierarchy affect larger areas such as cities and towns instead of the small sites where individual technologies are found. On a political dimension, governments make decisions at the top of the hierarchy, while mostly households and individuals make decisions at the bottom.

30 See L. Schipper, ed., *On the Rebound: The Interaction of Energy Efficiency, Energy Use, and Economic Activity,* special issue of *Energy Policy* 28, 6-7 (2000).

31 The other provinces are somewhere between these two. Hydropower provinces, such as British Columbia, Manitoba, and Newfoundland, are more similar to Quebec, and thermal provinces, such as Saskatchewan, New Brunswick, and Nova Scotia, are more similar to Alberta. With nuclear power, hydropower, and coal, Ontario is in between.

32 While the assumption that hydro does not emit GHGs has been generally agreed on in international negotiations, there is not universal agreement. Some research has found hydro reservoirs to emit more methane than would otherwise occur if the same area had remained in its previous unflooded state.

33 The implicit assumption behind the Kyoto Protocol is that the world realizes a net benefit from achieving the Kyoto target, although this does not mean that every country would benefit equally or even benefit at all from preventing climate change. We make no judgment here on the validity of the net benefit assumption. We simply note that, while the net global benefit of reducing global GHG emissions appears to be obvious to many researchers, there are those who point out that cost-benefit analysis may show the benefits to be lower than assumed and that the Kyoto agreement itself may not be beneficial. For detailed analysis, see W. Nordhaus, *Managing the Global Commons: The Economics of Climate*

Change (Cambridge, MA: MIT Press, 1994); and J. Weyant and J. Hill, eds., *The Costs of the Kyoto Protocol: A Multi-Model Evaluation*, special issue of *The Energy Journal* (1999).

34 See W. Nordhaus, "To Slow or Not to Slow: The Economics of the Greenhouse Effect," *Economic Journal* 101 (1991): 920-37; C. Carraro and J.-C. Hourcade, eds., *The Optimal Timing of Climate Change Policies,* special issue of *Energy Economics* 20, 5-6 (1998); and C. Azar, "The Timing of CO_2 Emissions Reductions: The Debate Revisited," *International Journal of Environment and Pollution* 10, 3-4 (1998): 508-21.

35 Weyant and Hill, eds., *The Costs of the Kyoto Protocol.*

36 CIMS originally stood for the Canadian Integrated Modelling System. However, as we are applying our model to other countries, we now just use the acronym CIMS.

Chapter 3: A Method for Estimating Policy Costs

1 For a description and contrast of the two models, see M. Jaccard, R. Loulou, A. Kanudia, J. Nyboer, A. Bailie, and M. Labriet, "Methodological Contrasts in Costing Greenhouse Gas Abatement: Optimization and Simulation Modeling of Microeconomic Effects in Canada," *European Journal of Operations Research,* in press.

2 The optimization approach has other important advantages, however. For example, an optimization model can simulate very complex integrated systems (multiple objectives, international trade) with greater ease than a simulation model. In the future such models may also be able to incorporate firm and household preferences in their parameters and decision algorithms; they have not been a focus in the past.

3 Informetrica, *Macro-Economic Impacts of GHG Reduction Options: National and Provincial Effects* (Ottawa: Informetrica, 2000); Department of Finance, Economic Studies and Policy Analysis Division, *A Computable General Equilibrium Analysis of Greenhouse Gas Reduction Paths and Scenarios* (Ottawa: Department of Finance, 2000).

4 While integrated modellers agree that simultaneous simulation is preferable, the AMG wanted to avoid the black-box syndrome of modelling, in which multiple interactive effects hinder interpretation. Given the relatively small size of the macroeconomic response (see Chapter 4), it does not appear that simultaneous simulation would have resulted in a significantly different outcome in terms of sectoral and regional allocation of reductions and costs, but in future modelling we intend to test this assumption.

5 The energy demand component of the model, previously called ISTUM, was first developed in the early 1980s by the US Department of Energy as an energy use model of the industrial sector.

6 In this respect CIMS resembles models developed and applied by the electric utility industry in the 1980s for estimating the effects of policies intended to influence technology choices for energy efficiency and fuel-switching objectives. CIMS has been used by electric and gas utilities in Canada for this purpose.

7 The model has been applied to testing a broad-based environmental tax policy that included GHG, water use, and solid waste taxes and recycling of the resulting tax revenue as reduced payroll charges to firms. See M. Jaccard and A. Taylor, "Simulating Environmental Tax Shifting: Developments in Economy-Materials Policy Modelling," working paper, Simon Fraser University, 2002.

8 A computable general equilibrium model is an example of a model that would score high on equilibrium feedbacks and would hopefully include consumer and firm preferences but is unlikely to include technology explicitness.

9 See I. Nystrom and C.-O. Wene, "Energy-Economy Linking in MARKAL-MACRO: Interplay of Nuclear, Conservation, and CO_2 Policies in Sweden," *International Journal of Environment and Pollution* 12, 2-3 (1999): 323-42.

10 For an overview of economy-climate integrated models, see C. Kolstad, "Integrated Assessment Modeling of Climate Change," in *Economics and Policy Issues in Climate Change,* ed. W. Nordhaus (Washington, DC: Resources for the Future, 1998), 263-87.

11 While not opposed to such modelling approaches, we caution that the usefulness of a model as a policy tool diminishes if the interactive effects and uncertainties are extended beyond the conceptual grasp of decision makers. An alternative to the one-big-model approach is to run two or more models in sequence. Doing so allows users to understand

what one model is showing before its outputs become the inputs for the next model in the sequence, what is sometimes referred to as *soft-linking*.

12 Issues in estimating these elasticities are similar to those for any macroeconomic model at a corresponding level of sectoral disaggregation.

13 There is considerable evidence that the pace of equipment turnover depends on the economic cycle, but over the long term, as simulated by CIMS, age is a fairly reliable and simple predictor.

14 Optimization modellers refer to this as the *reference energy system*.

15 While the CIMS industrial sector has great technological detail, its level of sectoral disaggregation is much less than for a typical macroeconomic model because a few energy-intensive sectors are represented in great detail, while the rest of the economy, including the entire service sector, is lumped into a single aggregate sector.

16 One of the few models operating at this level of technological and behavioural detail is the NEMS model of the US government. That data-intensive model requires a staff of about 40 for upkeep and application.

17 At the Energy and Materials Research Group, we operate (since 1992) one of the energy data centres funded primarily by the Canadian government, along with contributions from utilities and industry associations. Our centre is the Canadian Industrial Energy End-Use Data and Analysis Centre.

18 These four factors are also emphasized in R. Stavins, "The Costs of Carbon Sequestration: A Revealed Preference Approach," *American Economic Review* 89, 4 (1999): 994-1009.

19 For an example of using expert judgment in estimating GHG emission reduction, see A. Manne and R. Richels, "The Costs of Stabilizing Global CO_2 Emissions: A Probabilistic Analysis Based on Expert Judgments," *The Energy Journal* 15, 1 (1994): 31-56.

20 Revealed discount rates cover both of these factors because the new technologies of interest to energy-economy modellers are those that increase energy efficiency through irreversible long payback investments.

21 Sources are listed under the table. For an overview, see K. Train, "Discount Rates in Consumers' Energy-Related Decisions: A Review of the Literature," *Energy* 10, 12 (1985): 1243-54; and J. Nyboer, *Simulating Evolution of Technology: An Aid to Energy Policy Analysis: A Case Study of Strategies to Control Greenhouse Gases in Canada*, PhD diss., Simon Fraser University, 1997.

22 For an example, see D. Bunch, M. Bradley, T. Golob, R. Kitamura, and G. Occhiuzzo, "Demand for Clean-Fuel Vehicles in California: A Discrete-Choice Stated Preference Pilot Project," *Transportation Research* 27A, 3 (1993): 237-53. The values from this study have been updated to reflect changing technology options since 1993.

23 Because of the uncertainty surrounding the monetary value, we do this only in special circumstances. Without a discrete choice survey, it is difficult to determine if we have understood the decision-making process or even if the observed behaviour will endure over the 20-year simulation period.

24 Thus, the wider the variance, the lower the price elasticity of demand of the model. To compare the model's price responsiveness with those of other models (including the effect of discount rates and other behavioural parameters), the analyst can shock the model with a wide range of energy prices and then use the output (*pseudo-data*) for an econometric estimation of elasticity. See M. Jaccard and C. Bataille, "Estimating Future Elasticities of Substitution for the Rebound Debate," *Energy Policy* 28, 6-7 (2000): 451-55.

25 A. Grubler, N. Nakicenovic, and D. Victor, "Dynamics of Energy Technologies and Global Change," *Energy Policy* 27, 5 (1999): 247-80.

26 These service demand projections are indicated in terms of floor space (for the commercial and residential sectors), vehicle kilometres travelled (for transportation), and energy needs (for electricity and natural gas supply sectors).

27 Although the model was simulated to 2030, the AMG decided that only the period to 2022 would be considered for the estimation of policy costs.

Chapter 4: National Estimates

1 Further inducing measures means that a measure included in Path 0 would be more strongly applied. Thus, for example, if a measure causes half of consumers to buy high-efficiency

fridges, then a stronger application would influence even more consumers to buy efficient fridges. New actions are based on the technology options represented within CIMS.

2 Costs are low in Path 0, but it does not reach the target, so only Path 1 is an appropriate comparison.

3 Canadian economic output in the year 2000 is approximately 1 trillion dollars. Statistics Canada, CANSIM, Matrix 6547 and Catalogue 13-001-XIB. Last modified 1 February 2001.

4 The Informetrica simulations were based on the assumption that Canada would act alone. Because we are assuming a comparable GHG emission reduction effort by the United States and other industrialized trading partners, the macroeconomic impacts may be even less. However, because little of the impact in the Informetrica analysis results from changing trade patterns, the results reported here are an adequate approximation of macroeconomic effects.

5 GDP loss is compounded over time; Informetrica's reported annual impact of 2% means that the economy grows at a slower rate throughout the time period, and the loss would accumulate to a total decrease of economic output of about 3% by 2010 from the level at which it would otherwise be.

6 Mark MacKinnon, "Emission Cuts Not Seen Hurting Economy," *Globe and Mail*, 7 August 2000: B1, 2.

7 She is referring to a $30 billion loss in income in her speech. Nancy Hughes Anthony, "Climate Change: A National Issue Affecting Canadians," address notes, Joint Press Conference, Canadian Chamber of Commerce and the Canadian Association of Petroleum Producers, 4 March 2002, Calgary, AB. Retrieved 14 March 2001 from the Chamber of Commerce Web site <www.chamber.ca/newpaged/news.html>.

8 Informetrica, *Macro-Economic Impacts of GHG Reduction Options: National and Provincial Effects* (Ottawa: Informetrica, 2000), Tables 14-18, 43-44.

9 From a private perspective, the creation and use of a permit system would create a market-based return for GHG emission reductions, but there is no net financial return from the perspective of the entire economy. Nevertheless, increased investment in emission-reducing technologies does result in some savings in energy costs.

10 This assumes baseboard heating in a pre-1976 home, approximately 3,900 degree days of heating (typical of Toronto's climate), an electricity price increase from 8.4¢/kWh to 11.8¢/kWh (in 2010), and an annual service charge of $42. Annual residential heating costs will vary according to weather, occupant behaviour, and differences in heating systems and controls, as well as house size and vintage.

11 This assumes a standard efficiency furnace in a pre-1976 home, approximately 3,900 degree days of heating (typical of Toronto's climate), the natural gas rates described in the text, and an annual service charge of $92. Annual residential heating costs will vary according to weather, occupant behaviour, and differences in heating systems and controls, as well as house size and vintage.

12 Based on 9.7 litres/100 km (average efficiency of a new car in 1995) and an average annual distance driven of 21,580 km. Natural Resources Canada, *Canada's Energy Outlook* (Ottawa: NRC, 1997), 32-33. Gasoline prices are based on Ontario's 2010 reference case and national target prices.

13 R. Loulou et al., *Integrated Analysis of Options for GHG Reduction with Markal* (Montreal: HALOA, 2000). Prepared for the Analysis and Modelling Group, National Climate Change Process.

14 US Energy Information Administration, *Impacts of the Kyoto Protocol on US Energy Markets and Economic Activity* (Washington, DC: EIA, 1998).

15 Of course, the parameters in any model, whether top-down or bottom-up, can be adjusted to provide atypical results for that kind of model. In a recent application of the NEMS model, market-observed discount rates and other decision factors in technology adoption were adjusted in the model to reflect a high responsiveness to a mix of information and regulatory and fiscal policies. Also, only financial costs were used to estimate total costs. Interestingly, the application of NEMS still showed that, even with a wide array of policies and significant energy price increases (30% for gasoline), the United States would not achieve its Kyoto target of 7% below 1990 levels, ending up instead at 9% above them in 2010. However, total costs to the US economy were estimated to be negligible, unlike in the

earlier NEMS study and our own research for Canada with CIMS. See M. Brown, W. Short, and M. Levine, eds., "Scenarios for a Clean Energy Future," special issue of *Energy Policy* 29, 14 (2001).

16 The model has been applied in various published studies. The figures reported here are based on a recent recalibrated version of the model reported in a Natural Resources Canada memo. Paul Monfils, "Impacts of the Kyoto Protocol, Multi-Sector, Multi-Region Trade Model (MS-MRT)," Natural Resources Canada, February 2000.

17 We report the effects of alternative values for key assumptions. A more comprehensive sensitivity analysis would involve running a number of simulations with the values for all key uncertainties allowed to change simultaneously. Doing so would provide a distribution of model outcomes. If probabilities can be associated with the changed input values and assumptions, then a probability distribution of outcomes could be produced. This latter approach is known to practitioners as Monte Carlo simulation.

18 This implies higher costs for Canada than for the United States, but this seems plausible when one of the key differences between the two energy systems is the higher use of coal for electricity generation in the United States. While reducing emissions by switching away from coal is not inexpensive, it is generally low cost relative to options in the transportation sector, options that Canada would need to rely on to a greater degree.

19 R. Murphy and M. Jaccard, "Costing Greenhouse Gas Abatement in Canadian Transportation: Simulating Policies that Influence Preferences," working paper, Energy and Materials Research Group, Simon Fraser University, 2002.

20 These sensitivity numbers are based on runs conducted by the MARKAL modelling team using their model. We have estimated the relative effect to be about the same with a CIMS simulation.

21 See, for example, S. Holloway, S. Van der Straaten, and R. Van der Straaten, "The Joule II Project: The Underground Disposal of Carbon Dioxide," *Energy Conversion and Management* 36 (1995): 519-22.

22 This was estimated using the MARKAL model.

23 R. D'Abate, *Modelling Greenhouse Gas Abatement Strategy in the Canadian Electricity Sector* (Burnaby, BC: School of Resource and Environmental Management, Simon Fraser University, 2001), Report 282.

24 In the national process, a crude sensitivity was conducted in that both the TIMS model and the CaSGEM model were used to simulate the macroeconomic feedbacks from the microeconomic costs. However, the CaSGEM model did not apply directly the microeconomic cost information from CIMS or MARKAL.

25 Scenarios were developed by the Analysis and Modelling Group for the NCCP and tested on several models, although we interpret the definitions of the scenarios differently here. The permit prices represent the prices for GHG emission reduction credits on the world market, as estimated by the Analysis and Modelling Group, based on a US Department of Energy study. Permit prices varied over times: from \$24/t CO_2e in 2010 to \$35/t CO_2e in 2020, and from \$58/t CO_2e in 2010 to \$50/t CO_2e in 2020.

26 We are somewhat skeptical, however, of the substantial negative cost values (i.e., net benefits) indicated at the lower permit-trading prices. Some of the apparently low-cost actions provided by the Issue Tables lacked the full accounting of losses to consumers' value that are normally included in the CIMS model. Our subsequent research in the transportation sector, for example, suggests that measures to reduce vehicle use could be more costly than assumed in the data provided by the Issue Table. If these cost estimates are indeed too low, then the cost curve in Figure 4.7 should rise more quickly, although it would still reach 180 Mt at a price of \$120/t CO_2e, and the total costs in Figure 4.10 would not be negative at the permit-trading price of \$22/t CO_2e.

Chapter 5: Sectoral Estimates

1 Environment Canada, *Canada's Greenhouse Inventory, 1990-1998: Final Submission to the UNFCC Secretariat,* Volume 2 (Ottawa: Environment Canada, 2000).

2 Buildings Table, National Climate Change Process, *Buildings Table Options Report: Residential Sector* (Ottawa: Natural Resources Canada, 1999).

3 The household costs in Table 5.5 do not include vehicle costs as these are presented in the transportation sector.
4 Municipalities Table, National Climate Change Process, *Municipalities Table Options Paper: Final Report* (Ottawa: Natural Resources Canada, 1999).
5 See Chapter 2 for a description of cogeneration and district energy systems.
6 In 2000 the transportation sector consumed 2,280 PJ of energy, and total manufacturing energy consumption was 1,936 PJ. Statistics Canada, *Quarterly Report on Energy Supply-Demand in Canada 2000-IV* (Ottawa: Statistics Canada, 2002), Table 1B.
7 "Other Activities" includes commercial service vehicles, where transport is incidental to the service being provided.
8 The energy shares are for 1998. Calculations are based on information in Statistics Canada, *Quarterly Report on Energy Supply-Demand in Canada, 1998-IV* (Ottawa: Statistics Canada, 1999), Tables 17 and 18. Reported emissions are for 1998. Environment Canada, *Canada's Greenhouse Inventory, 1990-1998: Final Submission to the UNFCC Secretariat,* Volume 2 (Ottawa: Environment Canada, 2000), Table 1.
9 These costs are levelled capital costs (22-year period at 10% discount rate) plus fuel, operating, and maintenance costs.
10 Environment Canada, *Canada's Greenhouse Inventory, 1990-1998: Final Submission to the UNFCC Secretariat,* Volume 2 (Ottawa: Environment Canada, 2000), Summary Table 2, p. 45.
11 Canadian Industrial Energy End-Use Data and Analysis Centre, *Development of Intensity Energy Indicators for Industry 1990-2000* (Burnaby, BC: Simon Fraser University, 2001), 1.
12 Canadian Industrial Energy End-Use Data and Analysis Centre, *A Review of Energy Consumption in Canadian Oil Refineries and Upgraders: 1990, 1994 to 2000* (Burnaby, BC: Simon Fraser University, 2001).
13 In the production of adipic acid, a necessary precursor to the production of nylon, large quantities of N_2O are released. At significant cost, the company developed a method that, while it consumes more energy, reduces N_2O release to nil. N_2O has a CO_2 equivalency rating of 310. Thus, even a small reduction can have a significant effect on total emissions.
14 Environmental and Energy Systems Studies, *The Technology Menu for Efficient End-Use of Energy – Volume 1: Movement of Material* (Lund, Sweden: Lund University, 1989).

Chapter 6: Regional Estimates

1 The territories are included with British Columbia.
2 Because per capita emission reduction costs in the transportation sector are fairly even, the main reason for different regional distributions is due to the industrial sectors.
3 The Municipality Table actions lacked the full accounting of losses to consumers' value that are normally included in the CIMS model, for instance, changes in value from moving to higher density housing.
4 The survey reported that in Vancouver it takes an average of 60 minutes to travel to and from work, 70% of commuters use solely their cars, only 10% use transit, and 20% walk at least partway. In comparison, 61% of Torontonians reported that they use only their cars to get to work. Statistics Canada, *General Social Survey on Time Use,* 1992, as cited in Katherine Marshall, "Getting There," in *Perspectives on Labour and Income* (Ottawa: Statistics Canada, 1994), 17-22.
5 The residential sector is a major user of electricity. The share of electricity use by homes in British Columbia is higher than Canada's average because many homes use electricity for heating.
6 Alberta has a lower provincial sales tax on fuel – 9¢/litre (as of 1 January 1999) – compared with between 11¢ and 16¢/litre in most other regions.
7 The sector target yields few reductions with little policy incentive (as represented by the shadow prices of \$0/t CO_2e for residential and \$10/t CO_2e for commercial) required to reach the sector target.
8 As noted earlier, actions in the agriculture and forestry sectors were not simulated within the CIMS model but were instead added to the model results afterwards. Therefore, the nonfinancial preferences of consumers and producers are either ignored or at least not

incorporated to the same extent, which suggests that the estimates of low costs (or even benefits) may be overly optimistic.

9 Quebec, Ministry of Natural Resources, *Implementation of the United Nations Framework Convention on Climate Change – Quebec Action Plan* (Quebec: MNR, 2000).

10 For example, Quebec's iron and steel industry is based on electric arc furnaces rather than on basic oxygen furnaces of the integrated mills in Ontario.

11 Methane emissions vary depending on the type of mine – open pit or underground – as well as on its depth.

12 The region's underground mines have methane that is dense enough to permit capture to generate electricity.

Chapter 7: Domestic Policy Options

1 We apply a fairly standard definition of environmental sustainability: to allow present generations to achieve their goals without impairing, via environmental degradation, the opportunities for future generations to also achieve their goals.

2 Numerous works describe the special challenges of environmental policy making. The following are a selection from the much larger list: E. Loehman and M. Kilgour, eds., *Designing Institutions for Environmental and Resource Management* (Cheltenham, UK: Edward Elgar, 1998); M. Lutz, *Economics for the Common Good: Two Centuries of Social Economic Thought in the Humanistic Tradition* (London: Routledge, 1999); T. Prugh, *Natural Capital and Human Economic Survival*, 2nd ed. (Solomons, MD: International Society for Ecological Economics, 1999); M. Sagoff, *The Economy of the Earth: Philosophy, Law, and the Environment* (New York: Cambridge University Press, 1988); T. Sandler, *Global Challenges: An Approach to Environmental, Political, and Economic Problems* (New York: Cambridge University Press, 1997). For a Canadian perspective, see R. Boardman, ed., *Canadian Environmental Policy: Ecosystems, Politics, and Process* (Toronto: Oxford University Press, 1992); and M. Hessing and M. Howlett, *Canadian Natural Resource and Environmental Policy* (Vancouver: UBC Press, 1997).

3 See L. Pal, *Beyond Policy Analysis: Public Issue Management in Turbulent Times* (Scarborough: International Thomson Publishing, 1997); and Executive Resource Group, *Managing the Environment: A Review of Best Practices*, report to the Ontario Government, Toronto, 2001.

4 The terms *compulsion* and *obligation* are therefore less appropriate because they can suggest that one is acting in response to an internal demand, a personal feeling of obligation. By the term *compulsoriness,* in contrast, we mean that, as it increases in a policy instrument, the firm or household has less freedom of action because of externally imposed constraints.

5 Characterizing environmental policies in terms of compulsoriness is one of several possible criteria for comparison. Economists usually characterize policy instruments in terms of the extent to which market incentives are applied, while some policy analysts characterize policy instruments in terms of the extent to which legal prescriptions are included. See, for example, A. Hussen, *Principles of Environmental Economics* (London: Routledge, 2000).

6 The federal government has developed a number of programs to "lead by example": Federal Buildings Initiative, Federal Industrial Boiler Program, Federal House in Order Strategy, FleetWise Program, and Green Power Procurement Initiative.

7 In terms of technologies, for example, a common regulation requires a plant to install the best available control equipment.

8 A. Durning and Y. Bauman, *Tax Shift* (Seattle: Northwest Environment Watch, 1998).

9 G. Svendsen, C. Daugbjergand, and A. Pedersen, "Consumers, Industrialists, and the Political Economy of Green Taxation: CO_2 Taxation in OECD," *Energy Policy* 29, 6 (2001): 489-97.

10 Most jurisdictions have systems for beverage containers that involve payment of a deposit upon purchase and refund upon return of the empty container. A vehicle feebate is the combination of higher purchase taxes for high-pollution vehicles, with the revenue used to reduce the tax rates on low-pollution vehicles. A. Taylor, M. Jaccard, and N. Olewiler, *Environmental Tax Shift: A Discussion Paper for British Columbians,* Green Economy Secretariat, Government of British Columbia, Victoria, 1999.

11 A related, less comprehensive, approach is the extension of third-party liability laws and the strengthened legal standing of those who might wish to take civil action in pursuit of compensation for damages from those who are harming a common property resource.

12 Environmental taxes are sometimes referred to as the Pigouvian approach, named after the economist (Pigou) who first focused on pollution costs and fees. Market-oriented regulations are sometimes referred to as the Coasian approach, named after the economist (Coase) who developed the idea that allocating tradable pollution rights might be a legitimate and economically efficient approach to environmental policy. For an overview of these and similar approaches, see R. Kosobud and J. Zimmerman, eds., *Market Based Approaches to Environmental Policy* (New York: Van Nostrand Reinhold, 1997); M. Andersen and R.-U. Sprenger, eds., *Market-Based Instruments for Environmental Management* (Cheltenham, UK: Edward Elgar, 2000); and G. Svendsen, *Public Choice and Environmental Regulation: Tradable Permit Systems in the United States and CO_2 Taxation in Europe* (Cheltenham, UK: Edward Elgar, 1998).

13 In practice, several conditions need to be met for cost minimization, a critical one being the existence of competitive permit-trading markets.

14 R. Stavins, "What Can We Learn from the Grand Policy Experiment? Lessons from SO_2 Allowance Trading," *Journal of Economic Perspectives* 2, 12 (1998): 69-88.

15 T. Berry and M. Jaccard, "The Renewable Portfolio Standard: Design Considerations and an Implementation Survey," *Energy Policy* 29, 4 (2001): 263-77.

16 We use the term *VES* generically; different jurisdictions may use different terms. California has had separate authority to set vehicle emission standards since the 1970s, but the standards established in 1990 represent a significant leap in terms of technology expectations and flexibility mechanisms. The California Air Resources Board is one of six quasi-independent regulatory boards operating under the authority of the California Environmental Protection Agency. Details of its policies are found on its Web site <http://www.arb.ca.gov>.

17 See A. Pilkington, "The Fit and Misfit of Technological Capability: Responses to Vehicle Emission Regulation in the U.S.," *Technology Analysis and Strategic Management* 10, 2 (1998): 211-24. In addition to issues about the technological challenges of zero-emission vehicles, the focus on end use and local air pollution creates some challenges when contemplating VES as a GHG emission abatement policy. For example, the full-cycle emissions of the fuel used in a fuel cell- or battery-driven car need to be considered.

18 A. Grubler, N. Nakicenovic, and D. Victor, "Modeling Technological Change: Implications for the Global Environment," *Annual Review of Energy and Environment* 24 (1999): 545-69; C. Azar and H. Dowlatabadi, "A Review of Technical Change in Assessment of Climate Policy," *Annual Review of Energy and Environment* 24 (1999): 513-44.

19 California House, 2001-2002 session, AB 1058. An Act to add Section 43018.5 to the Health and Safety code, relating to air quality. As introduced in the House, retrieved online 16 March 2002 <www.assembly.ca.gov>. As of 18 March 2002 the bill had been approved by the California Assembly and was awaiting approval by the California Senate. If it is passed into law, the California Air Resources Board will be required to develop and adopt regulations by 2005 that achieve a maximum feasible reduction of carbon dioxide emitted by noncommercial cars, light trucks, and other vehicles.

20 Although in some cases government will move quickly and forcefully. Examples would be where there is a sudden awareness of substantial environmental and health risks from a highly toxic substance or where the safety of domestic water or food has been compromised.

21 Toronto has been active in pursuing GHG reductions and was the first municipality in the world to commit itself to reducing emissions by 20% from 1990 levels.

22 See P. Hawken, A. Lovins, and L. Lovins, *Natural Capitalism: Creating the Next Industrial Revolution* (Boston: Little, Brown, 1999); and B. Nattrass and M. Altomare, *The Natural Step for Business: Wealth, Ecology, and the Evolutionary Corporation* (Gabriola Island, BC: New Society, 1999).

23 In another instance, while there was no net economic or productive benefit, Dupont of Canada spent $15 million to install a catalytic converter that reduced nitrous oxide released by 93%, an equivalent of about 11 million tonnes of CO_2 annually. This action, in conjunction with other activities focused primarily on energy efficiency, reduced Dupont's output of GHGs by 85%.

24 K. Harrison points out that there is still much confusion over voluntary and mandatory programs and that we have almost no evidence of the effectiveness of voluntary programs.

K. Harrison, "Talking the Donkey: Cooperative Approaches to Environmental Protection," *Journal of Industrial Ecology* 2, 3 (1999): 51-72.

25 M. Bramley, *Greenhouse Gas Emissions from Industrial Companies in Canada: 1998* (Drayton Valley, AB: Pembina Institute, 2000).

26 Analysis and Modelling Group, National Climate Change Process, *Canada's Emissions Outlook: An Update* (Ottawa: NCCP, 1999), 43. The Office of Energy Efficiency of Natural Resources Canada is aware of the verification problem and is now making significant efforts to incorporate into the design and assessment of its efficiency programs some of the sound verification techniques developed in the electric utility industry in North America in the 1980s, a situation in which claims of efficiency gains had to be verified by regulatory boards before utilities were allowed to recover their expenses. But even in this rigorous climate, the effects of voluntary programs have always been extremely difficult to verify, especially the claim that program effects endure for a long time.

27 See L. Schipper, ed., *On the Rebound: The Interaction of Energy Efficiency, Energy Use, and Economic Activity,* special issue of *Energy Policy* 28, 6-7 (2000).

28 See D.S. Rothman and S.M. de Bruyn, eds., *Probing into the Environmental Kuznets Curve Hypothesis,* special issue of *Ecological Economics* 25, 2 (1998).

29 M. Porter and C. Van der Linde, "Green and Competitive: Ending the Stalemate," *Harvard Business Review* 73, 5 (1995): 120-34.

30 D. Violette, *Evaluation, Verification, and Performance Measurement of Energy Efficiency Programs* (Boulder: Hagler Bailly Consulting, 1996); M.R. Muller, T. Barnish, and P.P. Polomski, "On Decision Making Following an Industrial Energy Audit," *Proceedings of the 17th National Industrial Energy Technology Conference*, 8-10 April 1995, Houston (College Station, TX: Texas A&M University, 1995), 150-55. L. Berry, "A Review of the Market Penetration of U.S. Residential and Commercial Demand-Side Management Programmes," *Energy Policy* 21, 1 (1993): 53-67.

31 In 2000 the federal government announced an additional $1.1 billion to go toward responding to greenhouse gases and other areas of climate change – $625 million as part of its budget in the spring and an additional $500 million in the fall when the government tabled the First National Business Plan. Prior to this increase, Ottawa was spending about $200 million on climate change activities among a variety of government programs.

32 D. Violette, *Evaluating Greenhouse Gas Mitigation through DSM Projects: Lessons Learned from DSM Evaluation in the United States* (Boulder: Hagler Bailly Consulting, 1998).

33 Future research will need to explore the comparative performances of this type of policy with the other major categories of policy instruments. For an example of this kind of comparative analysis, see A. Jaffe and R. Stavins, "Dynamic Incentives of Environmental Regulations: The Effects of Alternative Policy Instruments on Technology Diffusion," *Journal of Environmental Economics and Management* 29 (1995): S43-63.

34 For example, an electricity efficiency initiative, even if profitable, may not be a profitable way of reducing GHG emissions if the electricity sector has also switched to nonemitting fuels at the same time.

35 We follow here the proposal of W. McKibben, M. Ross, and P. Wilcoxen, "Emissions Trading, Capital Flows, and the Kyoto Protocol," *The Energy Journal* 20 (1999 supp.): 287-334. The permits would be allocated to industry and all upstream energy producers, meaning that consumers would simply experience an increase in the price of energy. The effect is also similar to the proposal to apply a tradable permit system to industry and a GHG tax system to consumers, as detailed by G. Svendsen, *Public Choice and Environmental Regulation* (Cheltenham, UK: Edward Elgar, 1998). For a comparative analysis of the welfare performance under uncertainty of GHG taxes, tradable permits, and hybrid systems, see A. Pizer, *Prices vs. Quantities Revisited: The Case of Climate Change* (Washington, DC: Resources for the Future, 1997).

36 In effect, this is a convoluted form of environmental tax shifting but one that should not draw much reaction given the small amounts and the absence of a de facto tax.

Chapter 8: The Next Steps

1 There are other justifications for unilateral action by any one country such as Canada.

They include the ancillary environmental benefits from GHG reduction policies (e.g., reduced local air pollution) and possible strategic economic development benefits by being a technology leader. However, our focus in this work is on the value of cost information to decision making.

2 For an illustration of the value of strategies to reduce uncertainty, see W. Nordhaus and D. Popp, "What Is the Value of Scientific Knowledge? An Application to Global Warming Using the Price Model," *The Energy Journal* 18, 1 (1997): 1-45.

Glossary

action — A change in equipment acquisition, equipment use rate, lifestyle, or resource management practice that changes net GHG emissions from what they otherwise would be.

auctioning emission permits — Allocation of emission permits by auction. Alternative to **grandfathering emission permits.**

autonomous energy efficiency index (AEEI) — The rate at which the economy becomes more or less energy intensive as the price of energy stays constant relative to the prices of other inputs. Alternative to **elasticity of substitution.**

baseline assumption — An estimate of the trajectory that society is currently on in order to test policy impacts. Also referred to as **reference case**, **probable forecast**, and **business as usual**.

bottom-up analysis/model — Analysis that focuses on the apparent financial costs of technologies that, if widely deployed to meet the energy service needs of firms and households, would lead to dramatic reductions in GHG emissions. Alternative to **top-down analysis/model.**

building envelope/shell — Building structure, including foundation, frame, insulation, siding, windows, doors, and roof.

business as usual (BAU) — See **baseline assumption**.

cap and tradable permit system — A market-oriented regulation that sets a total emission limit, or cap, for whatever entity is being regulated: it could be several firms, a sector of the economy, a country, or the globe. Shares of this emission limit are allocated as tradable emission permits by some method (historical emissions, emissions auction) to participants (individuals, businesses, provinces, even countries depending on the scope of the program). The price at which permits are traded reflects the carbon price (financial signal, **marginal cost**, shadow price) of the last unit of emissions reduced. It should be equivalent to the emission tax rate that would achieve the same level of emission reductions.

carbon leakage/leakage — Refers to the relocation, to avoid costs, of industrial activity from countries with GHG emission reduction policies to countries without such policies. The net effect is that global emissions do not decline by as much as intended, hence the term carbon leakage or leakage.

clean development mechanism — A modified version of **joint implementation** that gives credit to signatories of the **Kyoto Protocol** for undertaking GHG emission reduction projects in developing countries.

CO_2 equivalency (CO_2e) Conversion of non-CO_2 greenhouse gases into a CO_2 equivalency based on their lifetime global warming potentials.

cogeneration The coproduction of useful heat (domestic, industrial process) and electricity. Cogeneration can achieve system energy efficiency in the 80% range, exceeding the efficiency of single-purpose thermal electricity generation units.

community energy management (CEM) A concerted effort to more closely integrate residential and commercial activities, as well as to connect nodes of higher urban density with transit infrastructure. CEM may result in reduced energy consumption and thus lower GHG emissions.

consumers' surplus (consumers' value) The difference between what consumers are willing to pay for a good or service and what they actually pay for it (market price). In other words, this is the premium that consumers would be willing to pay in addition to market price for a good or service.

cost incidence The allocation of policy costs among different sectors or regions of the economy.

cost-benefit analysis Identifying all of the incremental monetizable effects – positive and negative – from a particular policy or set of policies and summing these to arrive at a single monetary value for net cost or net benefit. Costs or benefits occurring in different time periods are made temporally equivalent via the **discount rate**.

cost-effectiveness analysis A restricted form of **cost-benefit analysis** where alternative means of achieving the same benefit are compared. Because the benefit of each alternative is identical, benefits need not be monetized **(monetization)** and the alternatives can be compared strictly on the basis of costs.

criteria air contaminants (CAC) These seven air pollutants are Total Particulate Matter, Particulate Matter 10, Particulate Matter 2.5, Carbon Monoxide, Nitrogen Oxide, Sulphur Dioxide, and Volatile Organic Compounds.

decomposition equation An equation that explains a term by breaking it into its component parts. In this book it refers to changes in energy use or GHG emissions as expressed in terms of changes in specific ratios such as the fossil fuel share of energy of a sector, the energy intensity of the sector, the share of the sector in the total economy, the total economic output per person, and the total population. In energy/environment analysis, this is sometimes referred to as the **Kaya Identity.**

direct emissions Emissions that are produced directly by the end-use technologies or activities in a sector. Alternative to **indirect emissions.**

discount rate A formula for converting values in different time periods into comparable units. Even when there is no inflation, values in different time periods need to be adjusted for comparison in order to take into account time preferences of consumers between current and future consumption and the productivity of invested capital.

distributed generation Generation of electricity that is situated at or near the point of consumption. Examples include fuel cells, microturbines, and photovoltaics.

district energy system A system of pipes that connects separate buildings for the distribution of hot water and steam for heating, or cold water for cooling, from a central location.

double dividend Term applied to GHG emission reduction policies that, first, induce economically efficient energy-related investments (profitable fuel switching and energy efficiency) and, second, generate revenues (from **GHG permit auctions** or **GHG taxes**) that are used to reduce distortions in the tax system.

eco-efficiency Environmental improvement that can be attained with an ethic of combining voluntarism with profit maximizing.

elasticity of substitution (ESUB) The sensitivity of the relative use of inputs (capital, labour, energy, and materials) to changes in their relative costs. A high elasticity implies a high degree of switching away from an input, such as energy, as its relative cost increases. Alternative to **autonomous energy efficiency index.**

energy intensity Energy use per unit of output (output could be an intermediate good or service, or final good or service).

energy service demand The demand for services from energy-using equipment, buildings, and infrastructure. In the residential and commercial sector, services include lighting, space heating, domestic water heating, appliances, and electronic equipment. In the industrial sector, services include process heating, lighting, motive force, and specific electricity applications. In the transportation sector, the energy service is mobility. Some services are indicated in energy units and some in physical units. Thus, examples from different sectors are tonnes of newsprint in industry, person-kilometres-travelled in transportation, and m^2 of heated floorspace in commercial.

environmental leveraging Maximizing the attainment of environmental objectives by allocating subsidies such that they only contribute that share of costs necessary to motivate an action. An example would be federal or provincial infrastructure subsidies to municipalities that require maximum environmental benefit per dollar of subsidy.

environmental tax shifting Levying environmental taxes and recycling all of the tax revenue as rebates to those who pay the taxes or as reductions in other taxes or charges (such as income taxes or payroll charges). An example is the **vehicle feebate.**

equi-marginal principle Total costs are minimized when each participant reduces emissions up to the point where the cost of the next unit of emission reduction (**marginal cost**) is more expensive than the next unit for someone else.

fugitive emissions Noncombustion releases of GHGs from the production processing, transmission, storage, and delivery of fossil fuels.

grandfathering emission permits Allocation of emission permits based on share of total emissions at some earlier period. Alternative to **auctioning emission permits.**

greenhouse gas (GHG) tax A tax on the unit GHG content of energy and other commodities. The level of the tax is set to induce changes that achieve a targeted level of emission reductions.

hard-linking Automatic communication and resolution between models. Alternative to **soft-linking.**

heat pump A technology that uses mechanical energy to transfer heat from a medium of lower temperature to a medium of higher temperature in order to cool or heat. Examples are an air conditioner, a fridge, and a heat pump space heater. Mediums to supply heating or cooling include air, water, and underground.

indirect emissions Emissions resulting from the production or conversion of energy (oil refining, natural gas processing, electricity generation). Changes in final energy demand by end-use technologies will be associated with changes in indirect emissions as the demand for energy production and conversion changes. Alternative to **direct emissions.**

indirect impacts The change in demand for the output from a sector of the economy as the sectors which it supplies are affected by a domestic policy or external development.

induced impacts Changes in employment, wage rates, and productivity caused by changes to personal consumption and business investment.

intangible or nonmonetary costs and benefits Costs and benefits not usually characterized in monetary terms. For the purposes of **cost-benefit analysis**, analysts may attempt **monetization**.

intermediate goods and services Goods and services supplied by one productive sector of the economy as inputs to another. An example would be glass sold to the automobile manufacturing industry.

joint implementation Method of reducing GHG emissions in one country with the help of another country. The benefits of the reduction are shared between the two countries, both of which are signatories of the **Kyoto Protocol**.

Kaya Identity See **decomposition equation.**

Kyoto Protocol An international agreement signed in December 1997 by 84 countries that commits signatories to reducing aggregate GHG emissions to 5.2% below 1990 levels by the period 2008-2012. Different countries were allotted different targets. The agreement does not come into effect until ratified by 55 countries responsible for 55% of the signatory countries' GHG emissions.

learning curves A nonlinear function that depicts the rate at which the cost of a new technology declines – toward an asymptote or minimum level – with increased commercialization.

macroeconomic model/effects Model that focuses on aggregate changes in economic activity, investment, employment, trade, prices, productivity, and government balances. Alternative to **microeconomic model/effects.**

marginal cost The cost of reducing emissions by one more unit.

market-oriented regulation A form of regulation in which the manner of participation is at the discretion of the firm or household. Examples are **cap and tradable permits**, **renewable portfolio standard**, and **vehicle emission standard**.

measure The combination of an **action** and the **policy instrument** that induced it.

microeconomic model/effects Model that focuses on the decisions of firms and households. Alternative to **macroeconomic model/effects.**

monetization The estimation of monetary values, positive or negative, for intangible values.

national target A target in which the Canadian economy must achieve economy-wide GHG reductions to 6% below 1990 levels, allocated according to the **equi-marginal principle** in order to maximize **cost-effectiveness**. Alternative to **sector target.**

optimization model A model that uses linear or dynamic programming to maximize or minimize an objective function subject to constraints. Such a model could inform users of what GHG emission reduction costs would be if all businesses and consumers focused on minimizing the net present value of financial costs, had complete information across time and space, and behaved identically. In contrast to a **simulation**

model, this type of model is considered normative rather than predictive in that it suggests how households and firms optimally would behave rather than how they are likely to behave.

option value The increase in expected net benefits resulting from the decision to delay an irreversible investment decision under uncertainty.

permit trading price The CO_2e (or GHG) permit-trading price that results from an emission **cap and tradable permit system.** The price is assumed to reflect the incremental cost of reducing emissions to the level of the cap or, conversely, the value of consuming the last unit of carbon when a cap has been imposed.

policy/policy instrument An effort by public authorities to bring about an **action**. Generic policy instruments include command-and-control regulations, information, voluntarism, taxes, subsidies, **environmental tax shifting**, and **market-oriented regulations**. See also **measure.**

probable forecast See **baseline assumption.**

radiative forcing Ability to retain heat, which refers to the greenhouse-causing potential of different greenhouse gases.

rebound effect An increase in energy consumption that offsets in part the reduction in energy consumption resulting from an energy efficiency action. This may occur because the efficiency action leads to higher income because it reduces the relative cost of the energy service.

reference case See **baseline assumption.**

renewable portfolio standard (RPS) A **market-oriented regulation** that requires producers of electricity to guarantee that a minimum percentage of their electricity is generated at designated, renewable energy facilities. Producers can trade among themselves to meet the aggregate requirement.

revealed preferences Historic market information on how consumers value certain technology attributes. Alternative to **stated preferences.**

sector target A target in which all sectors of the Canadian economy must achieve an identical percentage GHG reduction such that their total emissions in 2010 are 6% below 1990 levels. Alternative to **national target.**

sensitivity analysis An exercise in which one tests for the importance to modelling results of changes in key parameters, data inputs, or external assumptions.

sequestration The act of storing GHGs. Carbon sequestration may involve storage of carbon or CO_2 in plant matter, human structures, geological formations, or the bottom of the ocean. A source of storage is called a **sink.**

simulation model Model that tries to predict how firms and households will respond to policies intended to induce GHG emission reduction actions. In contrast to an **optimization model**, this type of model is considered predictive rather than normative in that it suggests how households and firms are likely to behave rather than how they optimally would behave.

sink A source of storage of GHGs. The act of storing is referred to as **sequestration.**

social discount rate The rate used to discount (the opposite of compound) the net returns in future time periods from collective or public investments in order to compare these with the initial investment. While there is much debate about how to determine the appropriate rate, it is agreed to be at or between (1) the rate at which consumers are willing to sacrifice consumption in the present for greater consumption in future and (2) the marginal, pre-tax rate of return on private investments (corrected for different levels of risk and inflation).

soft-linking Running two or more models in sequence where the transfer of information is not automatic. This allows users to understand what one model is showing before its outputs become the inputs for the next model in the sequence. Alternative to **hard-linking.**

stated preferences Consumer preferences that are discerned from surveys asking consumers to state their preferences for new and emerging technologies in terms of a monetary premium or discount. Alternative to **revealed preferences.**

top-down analysis/ model Analysis that focuses on aggregate relationships between inputs and outputs of the economy, and the interplay of these through price feedback mechanisms. The relative costs and uses of inputs like energy, with their associated GHG emissions, are portrayed in production functions for firms and consumption or utility functions for households. Alternative to **bottom-up analysis/model.**

total factor productivity The ratio over time of the sum of all input costs to the value of total output.

vehicle emission standard (VES) A **market-oriented regulation** that requires automobile manufacturers to guarantee that a minimum percentage of aggregate vehicle sales meet designated standards for emission levels. Manufacturers can trade among themselves to meet the aggregate target.

vehicle feebate A form of **environmental tax shifting** that levies higher purchase taxes on high-pollution vehicles, with all revenue used to reduce the purchase taxes on low-pollution vehicles.

Selected Bibliography

Analysis and Modelling Group, National Climate Change Process. *Canada's Emissions Outlook: An Update.* Ottawa: National Climate Change Process, 1999.

–. *An Assessment of the Economic and Environmental Implications for Canada of the Kyoto Protocol.* Ottawa: National Climate Change Process, 2000.

Andersen, M., and R.-U. Sprenger, eds. *Market-Based Instruments for Environmental Management.* Cheltenham, UK: Edward Elgar, 2000.

Awerbuch, S., and W. Deehan. "Do Consumers Discount the Future Correctly? A Market Based Valuation of Residential Fuel Switching." *Energy Policy* 23, 1 (1995): 57-67.

Azar, C. "The Timing of CO_2 Emissions Reductions: The Debate Revisited." *International Journal of Environment and Pollution* 10, 3-4 (1998): 508-21.

Azar, C., and H. Dowlatabadi. "A Review of Technical Change in Assessment of Climate Policy." *Annual Review of Energy and Environment* 24 (1999): 513-44.

Beausejour, Louis, Gordon Lenjosek, and Michael Smart. "A CGE Approach to Modelling Carbon Dioxide Emissions Control in Canada and the United States." *World Economy* 18, 3 (1995): 457.

Bernstam, M.S. *The Wealth of Nations and the Environment.* London: Institute of Economic Affairs, 1991.

Berry, L. "A Review of the Market Penetration of U.S. Residential and Commercial Demand-Side Management Programmes." *Energy Policy* 21, 1 (1993): 53-67.

Berry, T., and M. Jaccard. "The Renewable Portfolio Standard: Design Considerations and an Implementation Survey." *Energy Policy* 29, 4 (2001): 263-77.

Boardman, R., ed. *Canadian Environmental Policy: Ecosystems, Politics, and Process.* Toronto: Oxford University Press, 1992.

Bramley, M. *Greenhouse Gas Emissions from Industrial Companies in Canada: 1998.* Drayton Valley, AB: Pembina Institute, 2000.

Brown, M., M. Levine, J. Romm, A. Rosenfeld, and J. Koomey. "Engineering-Economic Studies of Energy Technologies to Reduce Greenhouse Gas Emissions: Opportunities and Challenges." *Annual Review of Energy and the Environment* 23 (1998): 287-385.

Bruntland, G.H., et al. *Our Common Future.* World Commission on Environment and Development. Oxford: Oxford University Press, 1987.

Bunch, D., M. Bradley, T. Golob, R. Kitamura, and G. Occhiuzzo. "Demand for Clean-Fuel Vehicles in California: A Discrete-Choice Stated Preference Pilot Project." *Transportation Research* 27A, 3 (1993): 237-53.

Carraro, C., and J.-C. Hourcade, eds. *The Optimal Timing of Climate Change Policies.* Special issue of *Energy Economics* 20, 5-6 (1998).

Claussen, E., and L. McNeilly. *Equity and Global Climate Change: The Complex Elements of Global Fairness.* Washington, DC: Pew Center on Global Climate Change, 1998.

Comeau, L. *Rational Energy Program: Update and Summary of Key Measures to the Year 2010.* Ottawa: Sierra Club of Canada, 1998.

DeCanio, S.J. "Barriers within Firms to Energy-Efficient Investments." *Energy Policy* 21, 9 (1993): 906-14.

Department of Finance, Economic Studies and Policy Analysis Division. *A Computable General Equilibrium Analysis of Greenhouse Gas Reduction Paths and Scenarios*. Ottawa: Department of Finance, 2000.

Dowlatabadi, H. "Sensitivity of Climate Change Mitigation Estimates to Assumptions about Technological Change." *Energy Economics* 20, 5-6 (1998): 473-94.

Durning, A., and Y. Bauman. *Tax Shift*. Seattle: Northwest Environment Watch, 1998.

Edmonds, J., R. Roop, and M. Scott. *Technology and the Economics of Climate Change Policy*. Arlington, VA: Pew Center on Global Climate Change, 2000.

Energy Research Group. *Cost Curve Estimations for Reducing CO_2 Emissions in Canada: An Analysis by Province and Sector*. Prepared for Natural Resources Canada. Burnaby: Simon Fraser University, 1998.

–. *Integration of GHG Emission Reduction Options Using CIMS*. Prepared for the Analysis and Modelling Group, National Climate Change Process. Vancouver: Energy Research Group/ M.K. Jaccard and Associates, 2000.

Executive Resource Group. "Managing the Environment: A Review of Best Practices." Prepared for the Government of Ontario, Toronto, 2001.

Goulder, L.H., and S.H. Schneider. "Induced Technological Change and the Attractiveness of CO_2 Abatement Policies." *Resource and Energy Economics* 21, 3-4 (1999): 211-54.

Gritsevskyi, A., and N. Nakicenovic. "Modelling Uncertainty of Induced Technological Change." *Energy Policy* 28, 13 (2000): 907-22.

Grubb, M. "Policy Modelling for Climate Change: The Missing Models." *Energy Policy* 21, 3 (1993): 203-08.

Grubb, M., J. Edmonds, P. ten Brink, and M. Morrison. "The Cost of Limiting Fossil-Fuel CO_2 Emissions: A Survey and Analysis." *Annual Review of Energy and Environment* 18 (1993): 397-478.

Grubler, A., N. Nakicenovic, and D. Victor. "Modeling Technological Change: Implications for the Global Environment." *Annual Review of Energy and Environment* 24 (1999): 545-69.

Guay, L. "Constructing a Response to Ecological Problems under Scientific Uncertainty: A Comparison of Acid Rain and Climate Change Policy in Canada." *Energy and Environment* 10, 6 (1999): 597-616.

Harvey, L.D. Danny, R. Torrie, and R. Skinner. "Achieving Ecologically Motivated Reductions of Canadian CO_2 Emissions." *Energy* 22, 7 (1997): 705-24.

Hausman, J. "Individual Discount Rates and the Purchase and Utilitization of Energy-Using Durables." *Bell Journal of Economics* 10, 1 (1979): 33-54.

Hawken, P., A. Lovins, and L. Lovins. *Natural Capitalism: Creating the Next Industrial Revolution*. Boston: Little, Brown, 1999.

Hessing, M., and M. Howlett. *Canadian Natural Resource and Environmental Policy*. Vancouver: UBC Press, 1997.

Holloway, S., and R. Van der Straaten. "The Joule II Project: The Underground Disposal of Carbon Dioxide." *Energy Conversion and Management* 36 (1995): 519-22.

Huntington, H., L. Schipper, and A. Sanstad. "Editors' Introduction." *Markets for Energy Efficiency*. Special issue of *Energy Policy* 22, 10 (1994): 795-97.

Informetrica. *Macro-Economic Impacts of GHG Reduction Options: National and Provincial Effects*. Ottawa: Informetrica, 2000.

Intergovernmental Panel on Climate Change. *IPCC Second Assessment Report: Climate Change 1995*. Cambridge: Cambridge University Press, 1996.

–. *Climate Change 2001, Third Assessment Report*. Cambridge: Cambridge University Press, 2001.

Jaccard, M., A. Bailie, and J. Nyboer. "CO_2 Emission Reduction Costs in the Residential Sector: Behavioral Parameters in a Bottom-Up Simulation Model." *The Energy Journal* 17, 4 (1996): 107-34.

Jaccard, M., and C. Bataille. "Estimating Future Elasticities of Substitution for the Rebound Debate." *Energy Policy* 28, 6-7 (2000): 451-55.

Jaccard, M., L. Failing, and T. Berry. "From Equipment to Infrastructure: Community Energy Management and Greenhouse Gas Emission Reduction." *Energy Policy* 25, 13 (1997): 1065-74.

Jaccard, M., R. Loulou, A. Kanudia, J. Nyboer, A. Bailie, and M. Labriet. "Methodological Contrasts in Costing Greenhouse Gas Abatement: Optimization and Simulation Modeling of Microeconomic Effects in Canada." *European Journal of Operations Research*, in press.

Jaccard, M., and D. Montgomery. "Costs of Reducing Greenhouse Gas Emissions in the United States and Canada." *Energy Policy* 24, 10-11 (1996): 889-98.

Jaccard, M., J. Nyboer, and A. Fogwill. "How Big Is the Electricity Conservation Potential in Industry?" *The Energy Journal* 14, 2 (1993): 139-56.

Jacobsen, H. "Integrating the Bottom-Up and Top-Down Approaches to Energy-Economy Modelling: The Case of Denmark." *Energy Economics* 20, 4 (1998): 443-61.

Jaffe, A., and R. Stavins. "The Energy-Efficiency Gap: What Does It Mean?" *Energy Policy* 22, 10 (1994): 804-10.

–. "Energy-Efficiency Investments and Public Policy." *The Energy Journal* 15, 2 (1994): 43-65.

–. "Dynamic Incentives of Environmental Regulations: The Effects of Alternative Policy Instruments on Technology Diffusion." *Journal of Environmental Economics and Management* 29 (1995): S43-63.

Kanudia A., and R. Loulou. "Advanced Bottom-up Modelling for National and Regional Energy Planning in Response to Climate Change." *International Journal of Environment and Pollution* 12, 2-3 (1999): 191-216.

Kolstad, C. "Integrated Assessment Modeling of Climate Change." In *Economics and Policy Issues in Climate Change*, ed. W. Nordhaus, 263-86. Washington, DC: Resources for the Future, 1998.

Koopmans, C., and D. te Velde. "Bridging the Energy Efficiency Gap: Using Bottom-Up Information in a Top-Down Energy Demand Model." *Energy Economics* 23, 1 (2001): 57-75.

Kosobud, R., and J. Zimmerman, eds. *Market Based Approaches to Environmental Policy.* New York: Van Nostrand Reinhold, 1997.

Loehman, E., and M. Kilgour, eds. *Designing Institutions for Environmental and Resource Management.* Cheltenham, UK: Edward Elgar, 1998.

Lohani, B.N., and A.M. Azini. "Barriers to Energy End-Use Efficiency." *Energy Policy* 20, 6 (1992): 533-45.

Loulou, R., et al. *Integrated Analysis of Options for GHG Reduction with MARKAL.* Prepared for the Analysis and Modelling Group, National Climate Change Process. Montreal: HALOA, 2000.

Lovins, A., and H. Lovins. "Least-Cost Climate Stabilization." *Annual Review of Energy and the Environment* 16 (1991): 433-531.

Lutz, M. *Economics for the Common Good: Two Centuries of Social Economic Thought in the Humanistic Tradition.* London: Routledge, 1999.

Manne, A.S., and R.G. Richels. "The Costs of Stabilizing Global CO_2 Emissions: A Probabilistic Analysis Based on Expert Judgments." *The Energy Journal* 15, 1 (1994): 31-56.

McKibben, W., M. Ross, and P. Wilcoxen. "Emissions Trading, Capital Flows, and the Kyoto Protocol." *The Energy Journal* 20 (1999 supp.): 287-334.

McKitrick, R. "Double Dividend Environmental Taxation and Canadian Carbon Emissions Control." *Canadian Public Policy* 23, 4 (1997): 417-34.

Metcalf, G. "Economics and Rational Conservation." *Energy Policy* 22, 10 (1994): 819-25.

National Climate Change Process. *Canada's First National Climate Change Business Plan.* Ottawa: National Climate Change Process, 2000.

National Round Table on the Environment and the Economy. *Canada's Options for a Domestic Greenhouse Gas Emissions Trading Program.* Ottawa: National Round Table on the Environment and the Economy, 1999.

Nattrass, B., and M. Altomare. *The Natural Step for Business: Wealth, Ecology, and the Evolutionary Corporation.* Gabriola Island, BC: New Society, 1999.

Natural Resources Canada, Climate Change Secretariat. *Government of Canada Action Plan 2000 on Climate Change.* Ottawa: Natural Resources Canada, 2000.

Nordhaus, W. "To Slow or Not to Slow: The Economics of the Greenhouse Effect." *Economic Journal* 101 (1991): 920-37.

–. *Managing the Global Commons: The Economics of Climate Change*. Cambridge, MA: MIT Press, 1994.

–, ed. *Economics and Policy Issues in Climate Change*. Washington, DC: Resources for the Future, 1998.

Nordhaus, W., and D. Popp. "What Is the Value of Scientific Knowledge? An Application to Global Warming Using the Price Model." *The Energy Journal* 18, 1 (1997): 1-45.

Norton, B., R. Constanza, and R. Bishop. "The Evolution of Preferences: Why 'Sovereign' Preferences May Not Lead to Sustainable Policies and What to Do about It." *Ecological Economics* 24, 2-3 (1998): 193-211.

Nyboer, J. *Simulating Evolution of Technology: An Aid to Energy Policy Analysis: A Case Study of Strategies to Control Greenhouse Gases in Canada*. PhD diss., Simon Fraser University, 1997.

Nystrom, I., and C.-O. Wene. "Energy-Economy Linking in MARKAL-MACRO: Interplay of Nuclear, Conservation, and CO_2 Policies in Sweden." *International Journal of Environment and Pollution* 12, 2-3 (1999): 323-42.

The Pembina Institute. *After Kyoto: Allocating Responsibility for Reducing Canada's Greenhouse Gas Emissions*. N.p.: Pembina Institute, Canadian Energy Research Institute, Conference Board of Canada, 1998.

Peters, I., F. Ackerman, and S. Bernow. "Economic Theory and Climate Change Policy." *Energy Policy* 27, 9 (1999): 501-04.

Pilkington, A. "The Fit and Misfit of Technological Capability: Responses to Vehicle Emission Regulation in the U.S." *Technology Analysis and Strategic Management* 10, 2 (1998): 211-24.

Pindyck, R. "Irreversibility, Uncertainty, and Investment." *Journal of Economic Literature* 29, 3 (1991): 1110-52.

Porter, M.E., and C. van der Linde. "Green and Competitive: Ending the Stalemate." *Harvard Business Review* 73, 5 (1995): 120-34.

Priddle, R. "The Energy Dimension of Climate Change." Parliamentary Group for Energy Studies, US Energy Information Administration, 4 June 1997.

Prugh, T. *Natural Capital and Human Economic Survival*, 2nd ed. Solomons, MD: International Society for Ecological Economics, 1999.

Repetto, R., and D. Austin. *The Cost of Climate Protection: A Guide for the Perplexed*. Washington, DC: World Resources Institute, 1997.

Robinson, J. "Of Maps and Territories: The Use and Abuse of Socioeconomic Modelling in Support of Decision-Making." *Technological Forecasting and Social Change* 42, 3 (1992): 147-64.

Robinson, J., M. Fraser, E. Haites, D. Harvey, M. Jaccard, A. Reinsch, and R. Torrie. *Canadian Options for Greenhouse Gas Emission Reduction (COGGER): Final Report of the COGGER Panel to the Canadian Global Change Program and the Canadian Climate Program Board*. Ottawa: Royal Society of Canada, 1993.

Rolfe, C. *Turning down the Heat: Emissions Trading and Canadian Implementation of the Kyoto Protocol*. Vancouver: West Coast Environmental Law Research Foundation, 1998.

Rollings-Magnusson, S., and R.C. Magnusson. "The Kyoto Protocol: Implications of a Flawed but Important Environmental Policy." *Canadian Public Policy* 26, 3 (2000): 347-60.

Rothman, D.S., and S.M. de Bruyn. "Introduction: Probing into the Environmental Kuznets Curve Hypothesis." *Ecological Economics* 25, 2 (1998): 143-46.

Sagoff, M. *The Economy of the Earth: Philosophy, Law, and the Environment*. New York: Cambridge University Press, 1988.

Sandler, T. *Global Challenges: An Approach to Environmental, Political, and Economic Problems*. New York: Cambridge University Press, 1997.

Sanstad, A., and R. Howarth. "Normal Markets, Market Imperfections, and Energy Efficiency." *Energy Policy* 22, 10 (1994): 811-18.

Scheraga, J. "Energy and the Environment: Something New under the Sun?" *Energy Policy* 22, 10 (1994): 798-803.

Schipper, L., ed. "On the Rebound: The Interaction of Energy Efficiency, Energy Use, and Economic Activity." *Energy Policy* 28, 6-7 (2000): 351-54.

Schwanen, Daniel. *A Cooler Approach: Tackling Canada's Commitments on Greenhouse Gas Emissions*. Toronto: C.D. Howe Institute, 2000.

Smith, H. *Canadian Federalism and International Environmental Policy Making: The Case of Climate Change*. Kingston: Institute of Intergovernmental Relations, Queen's University, 1998.

Stavins, R. "The Costs of Carbon Sequestration: A Revealed Preference Approach." *American Economic Review* 89, 4 (1999): 994-1009.

Sutherland, R. "The Economics of Energy Conservation Policy." *Energy Policy* 24, 4 (1996): 361-70.

Svendsen, G. *Public Choice and Environmental Regulation: Tradable Permit Systems in the United States and CO_2 Taxation in Europe*. Cheltenham, UK: Edward Elgar, 1998.

Svendsen, G., C. Daugbjerg, and A. Pedersen. "Consumers, Industrialists, and the Political Economy of Green Taxation: CO_2 Taxation in OECD." *Energy Policy* 29, 6 (2001): 489-97.

Taylor, A., M. Jaccard, and N. Olewiler. "Environmental Tax Shift: A Discussion Paper for British Columbians." Green Economy Secretariat, Government of British Columbia, Victoria, 1999.

Train, K. "Discount Rates in Consumers' Energy-Related Decisions: A Review of the Literature." *Energy* 10, 12 (1985): 1243-54.

–. *Qualitative Choice Analysis*. Cambridge, MA: MIT Press, 1986.

Train, K., and T. Atherton. "Rebates, Loans, and Customers' Choice of Appliance Efficiency Level: Combining Stated- and Revealed-Preference Data." *The Energy Journal* 16, 1 (1995): 55-69.

United Nations Development Programme. *World Energy Assessment: Energy and the Challenge of Sustainability*. New York: United Nations Development Programme, 2000.

United Nations Environment Programme. *The United Nations Framework Convention on Climate Change*. Geneva: Information Unit on Climate Change, United Nations Framework Convention on Climate Change, 1992.

–. *Kyoto Protocol to the United Nations Framework Convention on Climate Change*. Conference of the Parties, Third Session, 1998 (FCCC/CP/1997/L.7/Add.1).

US Energy Information Administration. *Impacts of the Kyoto Protocol on U.S. Energy Markets and Economic Activity*. Washington, DC: Energy Information Administration, 1998.

Violette, D. *Evaluation, Verification, and Performance Measurement of Energy Efficiency Programs*. Boulder: Hagler Bailly Consulting, 1996.

–. *Evaluating Greenhouse Gas Mitigation through DSM Projects: Lessons Learned from DSM Evaluation in the United States*. Boulder: Hagler Bailly Consulting, 1998.

Williams, A. "Low-Cost Strategies for Coping with CO_2 Limits." *The Energy Journal* 11, 4 (1990): 35-59.

Wilson, D., and J. Swisher. "Exploring the Gap: Top-Down Versus Bottom-Up Analysis of the Cost of Mitigating Global Warming." *Energy Policy* 21, 3 (1993): 249-63.

Working Group I, Intergovernmental Panel on Climate Change. *Climate Change 2001: The Scientific Basis, Third Assessment Report*. New York: Cambridge University Press, 2001.

Index

Note: Page numbers in **bold** indicate figures or tables.